There are no Aliens

R. Vandaal

Published by Blur Nation Media, 2025.

THERE ARE NO ALIENS

First edition. May 11, 2025.

ISBN: 979-8231905607

Written by R. Vandaal.

Table of Contents

There are No Aliens: A Critique of Extraterrestrialism

Introduction: Reimagining the Extraterrestrial Question

In the vast emptiness between stars, humanity has projected its deepest hopes and most primal fears. We have filled the cosmic void with imaginative conceptions of alien beings and civilizations—some benevolent and wise, others threatening and destructive—all of which reveal more about our own psychology, culture, and intellectual limitations than they do about what might actually exist beyond our small blue world.

This book does not claim that extraterrestrial life is impossible or nonexistent. Such a claim would be both scientifically unsound and philosophically presumptuous. Rather, *There are No Aliens* challenges what I term *extraterrestrialism*: the dominant paradigm through which contemporary science, popular culture, and public discourse conceptualize potential alien life. This paradigm is deeply rooted in anthropocentrism, shaped by Western scientific traditions, and constrained by the limits of human imagination.

The title is deliberately provocative. When I say "there are no aliens," I mean there are no aliens as we have come to imagine them—no tentacled beings piloting advanced spacecraft, no hyper-intelligent civilizations building Dyson spheres around distant stars, no carbon-based life forms evolving through natural selection on Earth-like planets. These are human projections, products of our specific evolutionary and cultural history. They tell us much about ourselves but offer limited insight into what might truly exist beyond our planetary boundaries.

The arguments in this book engage critically with multiple disciplines—astrobiology, astronomy, evolutionary biology, philosophy of science, cultural studies, and anthropology. I examine how our definitions of "life," our understanding of "intelligence," and our conceptions of technological progress all reflect terrestrial, human-centered biases that may blind us to radically different possibilities.

By deconstructing extraterrestrialism, we open ourselves to more nuanced, less anthropocentric ways of thinking about cosmic existence. We might begin to imagine forms of being so alien to our experience that they would be imperceptible or incomprehensible to us. We might recognize that our scientific methods and instruments, powerful as they are, remain extensions of human perception designed to confirm human models of reality.

This critical approach does not aim to diminish the importance of the search for extraterrestrial life or to discourage scientific inquiry into this profound question. Quite the opposite—it seeks to expand and enrich this search by highlighting its current limitations and suggesting new avenues of investigation and contemplation.

As we journey through the chapters that follow, I invite readers to temporarily suspend their familiar notions of alien life and to embrace the vertigo that comes with truly confronting the unknown. Perhaps in acknowledging the profound limits of our imagination, we might paradoxically move closer to understanding what—or who—might share our universe.

Major Theories of Extraterrestrial Existence and Their Critiques

The Drake Equation

Theory Overview: Formulated by Frank Drake in 1961, this equation attempts to estimate the number of active, communicative extraterrestrial civilizations in our galaxy by multiplying seven variables:

$N = R^* \times fp \times ne \times fl \times fi \times fc \times L$

Where:

- R^* = Rate of star formation
- fp = Fraction of stars with planets
- ne = Number of habitable planets per planetary system
- fl = Fraction of habitable planets where life emerges
- fi = Fraction of life-bearing planets that develop intelligence
- fc = Fraction of intelligent civilizations that develop detectable technology
- L = Length of time such civilizations release detectable signals

Critique: The Drake Equation illustrates how scientific frameworks can become circular arguments. We have reliable data for only the first two or three variables; the rest are essentially educated guesses based on our single example—Earth. The equation presupposes that:

1. Intelligence follows our evolutionary pathway

2. Technology develops along similar trajectories across species
3. Communication necessitates electromagnetic signals
4. Civilizations progress along a recognizable timeline

These assumptions reflect human experience, not universal truths. The equation's framework inadvertently creates a filter that only allows us to conceptualize civilizations similar to our own.

The Fermi Paradox

THEORY OVERVIEW: Physicist Enrico Fermi posed the question: If the universe is so vast and old, with billions of stars and planets, where is everybody? Given the age of the universe, even conservative estimates suggest intelligent life should be abundant and some civilizations should have colonized significant portions of the galaxy.

Critique: The paradox rests on several questionable assumptions:

1. **Colonization imperative:** Projects human expansionist tendencies onto all potential intelligent species.
2. **Detectability assumption:** Presumes advanced civilizations would be detectable by our rudimentary technology.
3. **Technological determinism:** Assumes technological development follows a linear, predictable path toward interstellar travel.
4. **Temporal myopia:** Our civilization has been capable of detecting radio signals for merely a century—an infinitesimal period in cosmic time.

The paradox may not be a paradox at all, but rather evidence of our limited imagination regarding how intelligence might manifest in the cosmos.

The Great Filter Hypothesis

THEORY OVERVIEW: Proposed by Robin Hanson, this solution to Fermi's Paradox suggests there must be one or more highly improbable evolutionary steps (filters) that prevent life from developing into interstellar civilizations. The question becomes whether this filter is behind us (e.g., the emergence of complex cells) or ahead of us (e.g., self-destruction through technology).

Critique: This hypothesis:

1. Treats civilization as having a singular, universal developmental trajectory
2. Presupposes that expansion and communication are inevitable goals
3. Imposes a Darwinian competitive framework on cosmic intelligence
4. Implicitly defines "success" as technological advancement and spatial expansion

The filter concept reveals our difficulty imagining forms of existence that don't prioritize growth, expansion, and dominance—values deeply embedded in Western capitalist societies.

The Rare Earth Hypothesis

THEORY OVERVIEW: Developed by Peter Ward and Donald Brownlee, this hypothesis argues that Earth-like planets with complex life are extremely rare due to the improbable combination of astronomical, geological, chemical, and biological factors that enabled life on Earth.

Critique: This theory:

1. Mistakes the conditions for Earth-like life for the conditions of all possible life
2. Exhibits sampling bias by treating Earth as the template for habitability
3. Underestimates the diversity of environments where life might emerge
4. Imposes terrestrial biochemistry as the universal standard

The hypothesis reflects scientific conservatism rather than imaginative inquiry—a reluctance to consider radically different biochemistries and ecological systems.

Panspermia

THEORY OVERVIEW: This theory suggests that life exists throughout the universe and is distributed by meteoroids, asteroids, and comets. Earth's life may have originated elsewhere and been seeded here through these cosmic vectors.

Critique: Panspermia:

1. Merely shifts the question of origins without addressing it
2. Still assumes life resembles terrestrial biochemistry
3. Privileges a narrative of cosmic interconnection that appeals to human desires
4. Often invokes directed panspermia (intentional seeding by advanced civilizations), which introduces teleology without evidence

The theory's appeal may lie in its resonance with ancient mythologies of celestial origins rather than its explanatory power.

The Zoo Hypothesis/Planetarium Hypothesis

THEORY OVERVIEW: These related theories suggest that extraterrestrial civilizations are deliberately avoiding contact with Earth, either observing us as we might observe animals in a zoo (Zoo Hypothesis) or creating an illusion of an uninhabited cosmos (Planetarium Hypothesis).

Critique: These hypotheses:

1. Anthropomorphize alien motivations and ethics
2. Project human conservation/research frameworks onto cosmic intelligence
3. Center humanity as especially interesting or worthy of observation
4. Imply a cosmic hierarchy reminiscent of religious cosmologies

The psychological appeal of these theories lies in preserving human specialness while acknowledging our technological immaturity—a secular version of divine observation.

Post-Biological Intelligence Theory

THEORY OVERVIEW: Advanced civilizations likely transition from biological to artificial or post-biological intelligence, potentially existing as uploaded consciousnesses, quantum computations, or other substrate-independent forms that would be unrecognizable and might not prioritize physical exploration or communication.

Critique: While more nuanced than some theories, this perspective:

1. Assumes a linear progression from biological to digital existence
2. Projects human technological trajectories onto all possible intelligences
3. Privileges information processing as the ultimate form of intelligence
4. Often ignores the possibility of biological paths that diverge radically from ours

This theory reveals our current AI-focused technological paradigm more than it illuminates universal evolutionary principles.

The Simulation Hypothesis

THEORY OVERVIEW: Proposed by philosopher Nick Bostrom, this hypothesis suggests that what we experience as reality is actually a computer simulation created by an advanced civilization. This could explain the apparent absence of alien life—it simply wasn't programmed into our simulation.

Critique: The simulation hypothesis:

1. Rests on computational metaphors specific to contemporary human technology
2. Creates an infinite regress problem (who simulates the simulators?)
3. Offers an unfalsifiable escape from the constraints of empirical science
4. Functions more as metaphysics than as scientific theory

This theory may be symptomatic of our increasingly virtual culture, projecting current technological trends onto the structure of reality itself.

Kardashev Scale and Technosignatures

THEORY OVERVIEW: Proposed by Soviet astronomer Nikolai Kardashev, this scale classifies civilizations by their energy utilization: Type I (planetary), Type II (stellar), and Type III (galactic). SETI searches focus on detecting the technosignatures such civilizations might produce.

Critique: This framework:

1. Defines advancement solely through energy consumption and manipulation
2. Projects industrial capitalism's values of resource exploitation into cosmic evolution
3. Assumes technological advancement follows a single, universal trajectory
4. Privileges environmentally invasive technologies as markers of intelligence

The scale reflects mid-20th century Soviet and American industrial ambitions more than it offers a universal model of advancement.

Transcension Hypothesis

THEORY OVERVIEW: Developed by John Smart, this hypothesis suggests that advanced civilizations don't expand outward but rather inward, achieving higher states of complexity and efficiency through miniaturization, virtualization, and exploration of inner space rather than outer space.

Critique: While addressing some limitations of expansionist models, this hypothesis:

1. Still assumes a singular developmental trajectory for all intelligences
2. Privileges information density and processing as the ultimate goals
3. Applies Moore's Law-type thinking beyond its empirical foundation
4. Often invokes singularity concepts that border on technological mysticism

This theory represents a more sophisticated anthropocentrism, exchanging expansionism for computational transcendence while still projecting human technological values.

The common thread through all these theories is their rootedness in human experience and values. Even our most imaginative scientific frameworks for extraterrestrial life reveal the limits of our conceptual architecture. A truly radical approach would acknowledge that alien existence might operate according to principles so foreign to human experience that our current theoretical frameworks cannot accommodate them—not just different answers, but entirely different questions.

Chapter 1: The Extraterrestrialist Framework

"We have met the alien, and it is us." — Attributed to various science fiction writers

The Human Cosmos

In 1997, the Hale-Bopp comet streaked across Earth's skies. For most viewers, it was simply a celestial spectacle—a rare opportunity to witness a cosmic traveler making its 2,380-year journey around our sun. But for the 39 members of the Heaven's Gate cult in San Diego, California, the comet was something else entirely: a sign that alien beings had arrived to transport their souls to the "Next Level." Their mass suicide, carried out in the belief that they would be reborn aboard an extraterrestrial spacecraft trailing behind the comet, stands as one of the most stark reminders of how deeply alien narratives can penetrate human consciousness.

This tragic event represents an extreme case of what I call *extraterrestrialism*—a pervasive set of assumptions, narratives, and conceptual frameworks through which humans imagine, theorize about, and search for extraterrestrial life. Extraterrestrialism is not merely the belief in aliens but the entire epistemological and ontological structure we have built around the possibility of their existence. It encompasses everything from scientific protocols for detecting biosignatures on distant exoplanets to Hollywood depictions of bug-eyed invaders, from theological debates about the spiritual status of extraterrestrials to the mathematics of the Drake Equation.

My purpose in this chapter is not to dismiss the search for extraterrestrial life as futile or misguided. Rather, I aim to excavate the hidden assumptions, cultural biases, and anthropocentric projections that shape this search at every level. By making these implicit frameworks explicit, we may begin to recognize the limits of our imagination and perhaps expand the horizons of what we might consider possible beyond our terrestrial experience.

From Ancient Pluralism to Modern SETI

THE NOTION THAT EARTH might not be unique in hosting life is ancient. Democritus, the pre-Socratic Greek philosopher who first proposed that all matter consists of indivisible particles called atoms, also postulated the existence of innumerable worlds, some inhabited by living beings. The Epicureans later developed this idea into a more systematic cosmic pluralism. In his poem *De Rerum Natura* (On the Nature of Things), the Roman philosopher Lucretius wrote:

> *"Nothing in the universe is unique and alone and therefore in other regions there must be other earths inhabited by different tribes of men and breeds of beasts."*

This early extraterrestrialism was not empirical but philosophical, emerging from atomistic materialism rather than observation. It served primarily to challenge religious cosmologies that placed Earth at the center of divine creation. Already we can identify a pattern that persists to this day: theories about extraterrestrial life often function as vehicles for terrestrial philosophical, theological, or political agendas.

The Copernican revolution, which displaced Earth from the center of the cosmos, gave new impetus to speculation about alien life. In

1584, Giordano Bruno—who would later be burned at the stake by the Inquisition—wrote that *"there are countless suns and countless earths all rotating around their suns in exactly the same way as the planets of our system."* He further declared that *"these worlds are all inhabited, just as is our own."* Bruno's vision was still not grounded in empirical evidence but represented an extension of the Copernican principle that Earth holds no special position in the universe.

The invention of the telescope in the early 17th century brought planetary bodies into closer view, triggering a new wave of speculation. When Giovanni Schiaparelli observed what appeared to be channels (later mistranslated as "canals") on Mars in 1877, it sparked decades of Martian fever. Percival Lowell, a wealthy American astronomer, built an observatory largely dedicated to studying these supposed Martian canals, which he believed were irrigation works constructed by an advanced civilization facing environmental collapse. Lowell's theories, though scientifically unfounded, captured the public imagination and influenced a generation of science fiction, including H.G. Wells' *The War of the Worlds* (1898).

The early 20th century brought radio astronomy, and with it, the first serious attempts to detect signals from extraterrestrial civilizations. In 1959, physicists Giuseppe Cocconi and Philip Morrison published a paper in *Nature* suggesting that intelligent beings might use radio waves for interstellar communication. The following year, astronomer Frank Drake conducted the first systematic search for such signals in Project Ozma. The Search for Extraterrestrial Intelligence (SETI) had begun.

Drake's approach represented a watershed moment in extraterrestrialism. For the first time, the question of alien life was framed in terms of empirical detection rather than philosophical speculation. Yet this apparent turn toward scientific objectivity masked a deeper continuity with earlier approaches: Drake's famous equation for estimating the number of communicative civilizations in our galaxy still relied heavily on assumptions about the nature, development, and behavior of extraterrestrial life—assumptions inevitably rooted in terrestrial experience.

The Birth of Modern Extraterrestrialism

WHAT DISTINGUISHES modern extraterrestrialism from earlier cosmic pluralism is its hybrid character—a fusion of scientific methodology, cultural imagination, institutional practices, and philosophical assumptions. This framework emerged fully in the post-World War II era, shaped by several converging factors:

1. **The Cold War space race**, which transformed outer space into a contested frontier and site of superpower competition
2. **The rise of Big Science**, with its massive funding structures and institutional frameworks
3. **The golden age of science fiction**, which popularized specific visions of alien life and interstellar contact
4. **The development of astrobiology** as a legitimate scientific discipline
5. **The UFO phenomenon**, which created new narratives of extraterrestrial presence
6. **The environmental movement**, which encouraged a planetary perspective
7. **The advent of powerful astronomical tools** like radio telescopes and space-based observatories

These factors did not simply facilitate the search for extraterrestrial life; they fundamentally shaped how that search was conceptualized and conducted. The instruments we built to detect alien life were designed to find specific signatures—radio waves at particular frequencies, atmospheric gases in particular combinations, liquid water in particular states. These detection parameters were not neutral or comprehensive but reflected our understanding of life based on Earth's single example.

Consider NASA's famous "follow the water" strategy for Mars exploration. This approach makes perfect sense given what we know about life on Earth, where water is indeed essential. But it also risks creating a circular logic: we look for life in ways that would detect Earth-like life, and when we don't find it, we conclude that life is absent—rather than considering that it might simply take forms we haven't imagined or can't detect with our Earth-calibrated instruments.

The emergence of extraterrestrialism as a dominant paradigm reflects what philosopher Thomas Kuhn would call "normal science"—research firmly based on past scientific achievements that a particular scientific community acknowledges as the foundation for its further practice. In the case of extraterrestrialism, these foundations include assumptions about the universality of carbon-based biochemistry, the evolutionary pathway toward intelligence, and the technological development trajectory toward radio communication.

Science Fiction and Scientific Imagination

A CRUCIAL FEATURE OF modern extraterrestrialism is the complex interplay between scientific research and science fiction. Far from being merely a popularization of scientific theories, science fiction actively shapes the conceptual frameworks through which scientists approach the question of alien life.

Consider the influential case of astronomer Carl Sagan, who straddled the worlds of hard science and popular culture. Sagan was both a serious researcher—publishing papers on the greenhouse effect of Venus's atmosphere and the potential for life on Jupiter—and a science fiction author whose novel *Contact* was adapted into a major Hollywood film. As the host of the PBS series *Cosmos*, he brought visions of extraterrestrial life to millions of viewers, including many who would later become scientists themselves.

Sagan's scientific work was informed by his imaginative capacity to envision alien environments, while his fiction was grounded in scientific plausibility. This cross-fertilization exemplifies what I call the "imagination-hypothesis feedback loop"—a process whereby fictional representations of aliens influence scientific hypotheses, which in turn shape new fictional representations.

This feedback loop is evident in countless examples:

- The "little green men" of early pulp science fiction influenced early SETI researchers' focus on intentional signals rather than technological byproducts.
- Arthur C. Clarke's depiction of a monolith on the Moon in *2001: A Space Odyssey* inspired actual scientific discussions about how to detect alien artifacts in our solar system.
- The representation of communication with aliens in *Close Encounters of the Third Kind* parallels actual protocols developed by SETI researchers for potential first contact scenarios.
- The "extremophiles" discovered in Earth's most hostile environments in the 1970s and 80s expanded science fiction's repertoire of potential alien biologies.

This is not to suggest that science fiction dictates scientific research agendas. Rather, both emerge from and contribute to a shared cultural imagination about what alien life might be like. The danger lies in mistaking this culturally conditioned imagination for an objective framework—in believing that our creative visions of aliens represent the full spectrum of possibilities rather than a narrow slice shaped by our evolutionary history and cultural context.

The Anthropocentric Foundations of Extraterrestrialism

AT ITS CORE, EXTRATERRESTRIALISM is inescapably anthropocentric. Even when we try to imagine radically different forms of life—silicon-based organisms, sentient clouds of gas, or hive minds—we do so with brains evolved on Earth to navigate terrestrial environments and social landscapes. Our most alien aliens are ultimately transformations or combinations of terrestrial templates.

This anthropocentrism operates at multiple levels:

Biological: We prioritize carbon-based, water-dependent life because that's what we are. Our scientific instruments are calibrated to detect biosignatures associated with Earth-like metabolism—oxygen, methane, carbon dioxide in particular ratios. Even when we consider "weird life," we tend to simply substitute elements (silicon for carbon) rather than imagining fundamentally different organizational principles.

Cognitive: Our models of alien intelligence almost invariably involve information processing systems analogous to human brains or computers. We struggle to conceptualize intelligence that doesn't involve symbolic representation, problem-solving, or technological manipulation—yet there's no reason to assume these features are universal rather than peculiar to our evolutionary path.

Technological: From the Kardashev Scale to the Drake Equation, we assume that technological development follows a predictable trajectory toward greater energy harnessing, computational capacity, and communicative reach. This reflects the particular history of human technological development, especially its acceleration in the industrial and digital eras.

Communicative: SETI focuses primarily on deliberate signals in the radio or optical spectrum—communication methods humans have developed relatively recently. The assumption that aliens would choose these channels reflects our own communicative priorities and technologies.

Motivational: We attribute to aliens motivations familiar from human experience—curiosity, exploration, resource acquisition, colonization, or cooperation. Yet these motivations emerge from our specific evolutionary history as social primates in competitive environments. Alien psychology, if such a concept even applies, might operate according to entirely different principles.

These anthropocentric assumptions are not errors to be corrected but inherent limitations of human cognition. We cannot think outside our own cognitive architecture. What we can do, however, is recognize these limitations and develop methodologies that attempt to compensate for them—approaches that deliberately seek to counter our anthropocentric biases.

UFOs: A Mirror for Human Concerns

NO DISCUSSION OF EXTRATERRESTRIALISM would be complete without addressing the UFO phenomenon, which serves as a perfect case study of how purported alien encounters function as cultural mirrors reflecting human preoccupations.

The modern UFO era is conventionally dated to Kenneth Arnold's 1947 sighting of "flying saucers" near Mount Rainier, followed shortly by the infamous Roswell incident. It is surely no coincidence that these events occurred at the dawn of the Cold War, as nuclear anxiety permeated American culture. The aliens of this era were often portrayed as either harbingers of doom warning humanity about nuclear self-destruction or as invaders employing technologies reminiscent of atomic weapons.

As the environmental movement gained momentum in the 1960s and 70s, UFO narratives shifted accordingly. Alien abduction accounts began to feature environmental themes—extraterrestrials concerned about human destruction of Earth's ecosystem or conducting breeding programs to preserve human genetic material in advance of ecological collapse.

The alleged government cover-up of alien contact, central to UFO mythology since Roswell, reflects Cold War paranoia about government secrecy but has evolved to incorporate contemporary anxieties about deep state conspiracies and corporate-governmental collusion. Each era projects its specific fears and hopes onto the blank screen of the unexplained aerial phenomenon.

Even the beings themselves—the iconic "grays" with their oversized heads, diminutive bodies, and enormous eyes—embody distinctly modern concerns. Their physical form, emphasizing brain capacity over bodily strength, reflects anxieties about technological evolution outpacing biological adaptation. Their lack of genitalia and apparent reproduction through technological means speaks to changing human relationships with sexuality and procreation in the age of reproductive technologies.

The UFO phenomenon, therefore, is less interesting as potential evidence of extraterrestrial visitation than as a complex cultural system through which humans process collective anxieties, articulate hopes for cosmic intervention, and negotiate changing relationships with technology. Whether or not any UFO has an extraterrestrial explanation (a question beyond the scope of this analysis), the *narratives* we construct around UFOs reveal the cultural workings of extraterrestrialism in its most explicit form.

Breaking the Frame

HAVING SKETCHED THE contours of extraterrestrialism as a paradigm, we can now begin to imagine alternatives. What would it mean to think about potential cosmic company outside this framework? How might we approach the question of extraterrestrial life without simply projecting human biology, psychology, and sociality onto the cosmos?

This is not merely a philosophical exercise but a practical question for scientific research. If our search methods are calibrated to detect only life that resembles us, we may be missing the vast majority of what's actually out there. By critically examining the assumptions embedded in our search strategies, we might develop more comprehensive approaches.

Several promising directions deserve further exploration:

1. **Agnostic biosignature research** that looks for atmospheric disequilibria without presuming specific chemical signatures of life
2. **Anomaly-focused observation** that prioritizes phenomena we can't explain over those we expect to find
3. **Indigenous and non-Western cosmologies** that

conceptualize non-human entities according to different ontological frameworks
4. **Artificial intelligence systems** trained to recognize patterns without human-imposed categories
5. **Arts-science collaborations** that use aesthetic practices to expand the imaginative capacity of scientific research

These approaches don't abandon scientific rigor but rather supplement it with methodological diversity. They recognize that the question of extraterrestrial life is not just scientific but philosophical, requiring us to examine how our ways of knowing might limit what we can know.

In the chapters that follow, we will examine specific dimensions of extraterrestrialism in greater detail—from biological assumptions to technological projections, from statistical reasoning to religious implications. Throughout, our goal will be not to dismiss the possibility of alien life but to expand our conception of what such life might be, and by doing so, to deepen our understanding of our own place in the cosmos.

For in the end, the search for alien intelligence is also inevitably a search for self-understanding. By attempting to recognize the truly other, we come to better recognize ourselves—our peculiarities, our contingencies, and perhaps even our unique value in a universe that may indeed be teeming with life, though perhaps not life as we have imagined it.

Chapter 2: A Matter of Biology? Nature and Process

"Life as we know it is only one way of being alive." — Lynn Margulis

The Carbon Chauvinism Problem

In 2010, NASA held a press conference that momentarily captivated the scientific world. Researchers announced the discovery of a bacterium, strain GFAJ-1, found in California's Mono Lake, that appeared to do something remarkable: substitute arsenic for phosphorus in its DNA backbone. If confirmed, this would have been revolutionary—the first known organism to use an alternative biochemistry for one of life's fundamental building blocks. Scientists speculated about the implications for alien life: if terrestrial organisms could perform such substitutions, how much more varied might extraterrestrial biochemistry be?

Within two years, however, other researchers had demonstrated that GFAJ-1 did not, in fact, incorporate arsenic into its DNA. It was simply highly tolerant of arsenic while still requiring phosphorus to function. The original findings were retracted, and a promising avenue for imagining alien biochemistry seemed to close.

This episode illustrates what Carl Sagan termed "carbon chauvinism"—our tendency to assume that all life must be carbon-based simply because all known life on Earth is. This assumption permeates astrobiology at every level, from the biosignatures we search for on exoplanets to the environments we deem habitable. It represents a fundamental constraint on our extraterrestrialist imagination, one that warrants critical examination.

The case for carbon-based life is compelling. Carbon forms more compounds than all other elements combined due to its unique ability to form stable covalent bonds with itself and many other elements. It creates long chains, rings, and complex three-dimensional structures essential for the diversity and functionality of biological molecules. Carbon compounds dissolve well in water, which itself possesses remarkable properties as a universal solvent and temperature buffer. Together, carbon and water seem ideally suited for the complex chemistry of life.

But this very suitability should give us pause. The fact that carbon works so well for life as we know it could be precisely why we've failed to imagine alternatives adequately. Our biochemical imagination is constrained by what evolutionary biologist Stephen Jay Gould called "the contingency of history"—the particular path that life on Earth happened to take. This path was neither inevitable nor necessarily universal.

Consider silicon, often proposed as an alternative to carbon. Like carbon, silicon can form four bonds and create complex molecules. Silicon-based life might use silanes instead of hydrocarbons and might employ liquid methane or liquid nitrogen as solvents rather than water. The temperatures conducive to such biochemistry would be far colder than those required for Earth life, perhaps making gas giant moons like Saturn's Titan potential habitats.

Yet our discussion of silicon-based life typically amounts to simply substituting one element for another while retaining essentially Earth-like biological processes: metabolism, reproduction, evolution by natural selection. This represents a failure of imagination, not a genuine attempt to conceptualize alien biochemistry. We replace the carbon in our existing model with silicon but leave the model's fundamental structure intact.

More radical possibilities exist. Life might emerge based on:

- **Boron chemistry** in non-aqueous solvents like liquid ammonia
- **Sulfur-based biochemistry** in extremely hot environments
- **Metal-sulfide-based systems** that use redox coupling instead of carbon chemistry
- **Plasma-based organizations** in stellar atmospheres
- **Quantum-mechanical phenomena** at interfaces between different states of matter

These speculations push the boundaries of conventional biochemistry, but they still operate within a recognizable scientific framework. What about possibilities that challenge our very conception of life? What about entities organized according to principles we haven't yet discovered or cannot comprehend with human cognition?

Life as a Category Problem

THE DIFFICULTY IN IMAGINING alternative biochemistries points to a deeper issue: the category of "life" itself may be a terrestrially-bound concept rather than a universal one. When astrobiologists attempt to define life, they typically enumerate characteristics derived from Earth's biosphere: metabolism, homeostasis, organization, growth, adaptation, response to stimuli, reproduction. But these characteristics constitute a family resemblance pattern rather than necessary and sufficient conditions.

Some entities we consider alive exhibit only some of these characteristics. Viruses reproduce but lack independent metabolism. Sterile hybrid animals cannot reproduce but are clearly alive. Dormant seeds show no detectable metabolism for years yet remain viable. The boundaries of life are fuzzy even on Earth, where we share a common biochemical heritage.

This categorical ambiguity becomes more acute when we consider potential borderline cases in our own solar system. Are the complex organic molecules in Titan's hydrocarbon lakes precursors to life, early life, or something else entirely? Does the potential subsurface ocean of Europa harbor organisms we would recognize as alive, or entities organized according to principles that don't map neatly onto our categories?

The problem intensifies when we consider theoretical entities that might exist elsewhere in the cosmos. Imagine:

- Self-organizing patterns in interstellar plasma that maintain their structure for millions of years
- Information systems encoded in gravitational wave patterns
- Quantum entanglement networks that process and store information without discrete physical bodies
- Chemical gradients in gas giant atmospheres that maintain, replicate, and evolve their patterns over time
- Geological processes that exhibit inheritance, variation, and selection on timescales of billions of years

Would these entities be "alive" according to our definition? Would they be detectable by our instruments? Would we recognize them as separate entities at all, rather than as background processes or physical anomalies?

The philosopher Ludwig Wittgenstein famously argued that many conceptual problems arise from the misuse of language—from applying terms beyond their proper contexts. Perhaps "life" is such a term—useful within the terrestrial context where it evolved but increasingly strained when extended to cosmic scales and alien environments. Perhaps we need not a broader definition of life but a recognition that "life" itself is a local concept, a way of carving up the world that reflects our particular evolutionary and cultural history rather than a universal category.

The Observer Problem

OUR STRUGGLE TO DEFINE life objectively points to what physicists and philosophers call "the observer problem"—the inescapable influence of the observer on the observed. This problem is well-recognized in quantum mechanics, where the act of measurement affects the system being measured, but it manifests in more subtle ways in our search for extraterrestrial life.

Consider how we observe potential signs of life on other worlds. We use instruments calibrated to detect signatures familiar from Earth: atmospheric oxygen, methane in disequilibrium, chlorophyll absorption bands. These instruments are extensions of human perception, designed to detect patterns we associate with life as we know it. They are not neutral windows onto reality but specialized tools that highlight certain aspects of reality while remaining blind to others.

This observational bias operates at multiple levels:

Technological: Our instruments have specific detection ranges and sensitivities that privilege certain phenomena over others. Radio telescopes search in frequency bands we consider "logical" based on human communication priorities and physical constants, potentially missing signals in unexpected parts of the spectrum.

Temporal: Earth life operates on timescales we can comprehend—from the milliseconds of neural firing to the millennia of evolution. But what about processes too slow or too fast for our perception or attention? A geological intelligence might communicate over millions of years, its thoughts expressed through continental movements imperceptible in a human lifetime.

Scalar: We prioritize phenomena at scales relevant to human experience—roughly from microscopic to planetary. But what about organization at quantum scales or across interstellar distances? We might be embedded within vast cosmic intelligences without recognizing them, like cells unaware they are part of a multicellular organism.

Sensory: Human perception evolved to detect a narrow band of electromagnetic radiation (visible light), certain chemical signals, mechanical vibrations within specific frequencies, and so on. Entities that communicate through modalities beyond our sensory range might remain entirely imperceptible without specialized instruments.

Even with advanced technology extending our perceptual range, we interpret the data through cognitive frameworks shaped by evolution on Earth. Our pattern recognition systems evolved to identify relevant features in terrestrial environments—predators, prey, potential mates, edible plants. These systems may be fundamentally unsuited to recognizing truly alien patterns of organization.

The astrobiologist Sara Walker suggests that we need a new approach to life detection—one that focuses on identifying complex, non-equilibrium systems capable of information processing and causal influence rather than specific biochemical signatures. This "agnostic biosignature" approach might help transcend some of our observational biases, but it still operates within human cognitive frameworks. We may need to acknowledge that some forms of extraterrestrial organization will remain invisible to us not because of technological limitations but because of the inherent structure of human cognition.

Nature as Human Construct

THE OBSERVER PROBLEM in astrobiology connects to broader philosophical questions about "nature" as a category. Environmental philosophers and anthropologists have long argued that what we call "nature" is not simply the objective world out there but a culturally constructed concept with a specific history and function in Western thought.

The very distinction between "natural" and "artificial" emerges from a particular intellectual tradition that separates humans from the rest of the world—a separation reinforced during the Scientific Revolution and Enlightenment. Indigenous and non-Western traditions often conceptualize human-environment relationships differently, seeing humans as integral parts of natural systems rather than external observers or managers.

When we search for "natural" signs of life on other worlds, we import this cultural framework into our scientific methodology. We look for what we consider "natural" biosignatures—gases produced by metabolism, patterns created by photosynthesis—while remaining vigilant against "false positives" that might actually be abiotic processes. This dichotomy presupposes a clear distinction between the biological and the geological, the living and the non-living, that may not apply in alien contexts.

Consider Saturn's moon Enceladus, where complex organic compounds have been detected in plumes erupting from a subsurface ocean. Are these compounds "natural" in the sense of being produced by biological processes, or "natural" in the sense of resulting from abiotic chemistry? The ambiguity here is not merely empirical—a question to be resolved with better data—but conceptual. The categories themselves may be inadequate for what we're discovering.

The social construction of nature also appears in how we imagine alien environments. Science fiction typically portrays alien worlds as analogs to terrestrial biomes—forests, deserts, oceans—populated by creatures fulfilling ecological roles familiar from Earth. Even when these environments are extremely harsh by human standards (like the sulfuric acid clouds of Venus), we conceptualize them through terrestrial analogies. This is not merely a failure of imagination but a reflection of how deeply Earth-bound categories structure our thinking.

Breaking free from these constraints requires more than simply imagining more exotic environments. It requires recognizing the contingency of our conceptual frameworks themselves—acknowledging that our ways of carving up the world into "natural" and "artificial," "living" and "non-living," "biological" and "geological" reflect a specific intellectual history rather than universal categories. Alternative frameworks might yield entirely different ways of conceptualizing what exists beyond Earth.

Evolution Beyond Predation

PERHAPS THE MOST PROFOUND anthropocentrism in our extraterrestrialist imagination involves projecting Earth's evolutionary history onto alien worlds. We assume that natural selection—competition for resources, predator-prey relationships, adaptation to changing environments—must shape life everywhere as it has shaped life on Earth.

This assumption appears most clearly in how we imagine alien behavior and society. Science fiction routinely portrays aliens as either predatory invaders or advanced altruists who have transcended their evolutionary heritage—both projections of human evolutionary psychology rather than genuinely alien possibilities. Even serious scientific speculation about alien civilizations often assumes familiar evolutionary trajectories, with intelligence emerging as an adaptation for social competition or environmental manipulation.

But what if the fundamental dynamics of evolution operate differently elsewhere? What if, as the biologist Lynn Margulis argued, cooperation and symbiosis play a more central role than competition? Margulis demonstrated that major evolutionary transitions on Earth, including the emergence of complex cells, involved symbiotic mergers rather than competitive victories. Perhaps on worlds with different environmental pressures, such cooperative processes might dominate.

Consider a world where energy is abundant but complex organic compounds are scarce. Life there might evolve elaborate resource-sharing networks rather than competitive strategies, with natural selection favoring organisms that contribute to collective metabolism rather than those that optimize individual fitness. Or imagine a planet with extreme but predictable environmental fluctuations, where evolutionary pressure might favor suspended animation or information preservation over continuous activity and reproduction.

Even more radically, what if some environments allow for forms of inheritance and variation that don't involve the replication of discrete organisms at all? Information might be preserved and modified through environmental persistence rather than genetic encoding, creating evolutionary processes without recognizable "organisms" as we conceive them.

The evolutionary biologist Richard Dawkins introduced the concept of "extended phenotype" to describe how genes influence the world beyond the body of the individual organism—as when beaver dams or bird nests become expressions of genetic information. Taking this concept further, might there be entities whose "bodies" are more environmental than organismal, distributed through a medium rather than bounded by membranes?

Such speculations may seem fanciful, but they serve an important purpose: they help us recognize how thoroughly our conception of life is shaped by Earth's particular evolutionary history. The predatory, competitive aspects of evolution that have so captivated human imagination—the "survival of the fittest" narrative that has been misused to justify various social hierarchies—may be contingent rather than necessary features of how complex systems evolve.

By imagining evolutionary processes beyond predation and individual competition, we might open new avenues for detecting and recognizing alien life. We might look for signs of distributed information processing, for unusual resource cycling patterns, for persistence mechanisms that don't involve discrete reproduction. We might develop models of non-Darwinian or extended Darwinian evolution that help us recognize patterns we would otherwise miss.

Life Beyond Movement

ANOTHER PROFOUND CONSTRAINT on our extraterrestrialist imagination is our assumption that life must involve movement, consumption, and waste production. Earth life is metabolic—it takes in energy and materials, transforms them, and excretes waste products. This metabolism typically involves movement, whether at the cellular level (as with bacterial flagella) or the organismal level (as with animal locomotion).

What if life elsewhere does not move, consume, or excrete in ways we would recognize? What if it operates according to principles that don't involve the familiar metabolic processes of Earth life?

Imagine entities that:

- **Harvest energy** through quantum coherence rather than chemical reactions
- **Exchange information** but not matter with their environments
- **Process information** as their primary activity rather than physical maintenance
- **Exist as standing waves** in various media rather than as discrete bodies
- **Grow by accretion** rather than internal cellular division
- **Transform without moving** through internal reorganization rather than external navigation

Such possibilities challenge our intuitive sense of what constitutes a living entity. We are adapted to recognize movement as a primary sign of life—our visual systems are highly attuned to detecting motion in our environment, an adaptation with clear survival value for creatures that must either catch moving prey or avoid becoming prey themselves. This perceptual bias influences how we search for life: we look for changes over time, for dynamics rather than statics.

But what if some forms of extraterrestrial organization are primarily static from our perspective? What if they change so slowly or so quickly that their movements are imperceptible to human observation? What if they exist primarily as structures rather than processes, or as processes so different from familiar biochemical ones that we lack the conceptual framework to recognize them?

The physicist Freeman Dyson speculated about "eternal life" scenarios where intelligence might persist in an ever-cooling universe by slowing down its computational processes to conserve energy—thinking ever more slowly as the available energy diminishes. Such an entity might appear completely static to observation on human timescales, yet harbor complex internal dynamics when viewed at the appropriate temporal scale.

Conversely, processes that unfold too quickly for human perception might appear as steady states rather than dynamic systems. Quantum effects that collapse almost instantaneously, plasma formations that reconfigure at near-light speed, or computational processes that cycle through states at femtosecond intervals would appear to us as unchanging phenomena rather than as the activities of organized entities.

These considerations suggest that our search for extraterrestrial life needs to be temporally agnostic—aware that organization might exist at timescales radically different from those we associate with terrestrial biology. We need observational strategies that can detect both extremely slow changes (through long-term monitoring or geological records) and extremely fast ones (through high-speed instrumentation and computational analysis).

Toward a Post-Biological Astrobiology

THE CRITIQUES DEVELOPED in this chapter point toward what might be called a "post-biological" approach to astrobiology—one that seeks to transcend the limitations of Earth-based definitions of life without abandoning the scientific search for complex organization beyond Earth.

Such an approach would:

1. **Prioritize pattern detection over specific biosignatures**, looking for anomalous organizations of matter and energy without presupposing familiar biological processes
2. **Develop multiple, overlapping definitions of life** rather than seeking a single comprehensive definition, recognizing that different conceptual frameworks might reveal different phenomena
3. **Incorporate insights from complex systems theory**, information theory, and quantum mechanics to expand our conception of what constitutes an organized entity
4. **Engage seriously with non-Western and Indigenous perspectives** on the boundaries between living and non-living, human and non-human
5. **Use artistic and speculative methodologies** alongside standard scientific ones to expand the imaginative capacity of astrobiological research
6. **Develop new observational tools explicitly designed** to detect non-Earth-like patterns of organization
7. **Create simulation environments** that allow for the emergence of organizational principles different from those that govern Earth life

This post-biological astrobiology would not abandon the search for familiar life forms—carbon-based organisms using water as a solvent, engaging in metabolism and reproduction similar to Earth life. Such organisms may indeed exist and may even be common in environments similar to Earth's. But it would recognize that such life forms represent only one possibility among many, and that our search strategies should be as diverse as the possible forms of extraterrestrial organization.

The fundamental question shifts from "Is there life out there?" to "What forms of organization exist beyond Earth, and how might we recognize them?" This reframing acknowledges the anthropocentric limitations of our current approach while opening new avenues for exploration and discovery.

In the following chapters, we will examine other dimensions of extraterrestrialism—from probability calculations to technological projections, from religious implications to sociological functions. Throughout, we will continue to interrogate the human-centered assumptions that shape our cosmic imagination and limit our capacity to recognize what might truly be out there. For the greatest discovery in our search for alien life may not be the aliens themselves but the recognition of how thoroughly our vision of them has been shaped by our terrestrial experience—a recognition that might, paradoxically, allow us to see beyond that experience toward a more expansive cosmic perspective.

Chapter 3: Probabilities as Conclusions: Solving Fermi's Paradox

"Sometimes I think we're alone in the universe, and sometimes I think we're not. In either case, the idea is quite staggering." — Arthur C. Clarke

The Paradox That Isn't

In the summer of 1950, during a lunch at Los Alamos National Laboratory, the physicist Enrico Fermi posed a deceptively simple question: "Where is everybody?" His colleagues immediately understood the context—if the universe is as vast and ancient as we believe, with billions of stars much older than our sun, then intelligent life should have arisen many times over. Some of these civilizations should have developed interstellar travel, colonized large portions of the galaxy, and made their presence known to us. Yet we observe no evidence of their existence. This apparent contradiction between high probability calculations and observational absence has come to be known as Fermi's Paradox.

The paradox has spawned an entire subfield of speculative science, with proposed solutions ranging from the mundane (interstellar travel is physically impossible) to the catastrophic (advanced civilizations inevitably destroy themselves) to the conspiratorial (aliens are here but hiding from us). These proposals are generally presented as logical deductions from empirical facts, mathematical probabilities, and reasonable assumptions. But what if the paradox itself is not a logical problem to be solved but a symptom of our anthropocentric projection?

In this chapter, I argue that Fermi's Paradox exists primarily as an artifact of how we frame the question of extraterrestrial intelligence. It emerges from a series of questionable assumptions about how intelligence develops, what motivates intelligent beings, and what constitutes evidence of their existence. By examining these assumptions, we can understand why the "paradox" may not be paradoxical at all—and in the process, develop a more nuanced understanding of our place in the cosmos.

The Numbers Game: Drake's Equation and Statistical Overreach

ANY DISCUSSION OF FERMI'S Paradox inevitably involves the Drake Equation, formulated by astronomer Frank Drake in 1961 as a framework for estimating the number of active, communicative extraterrestrial civilizations in our galaxy. The equation multiplies a series of factors:

$N = R^* \times fp \times ne \times fl \times fi \times fc \times L$

Where:

- R^* = Average rate of star formation in our galaxy
- fp = Fraction of those stars with planets
- ne = Average number of habitable planets per star with planets
- fl = Fraction of habitable planets that develop life
- fi = Fraction of life-bearing planets that develop intelligence
- fc = Fraction of intelligent civilizations that develop detectable technology
- L = Length of time such civilizations release detectable signals

When Drake first proposed this equation at the Green Bank Conference, he and his colleagues estimated that there might be between 1,000 and 100,000,000 communicative civilizations in our galaxy alone. Modern estimates vary widely but often yield large numbers that make Earth's apparent isolation seem puzzling or improbable.

The problem, however, lies not in the equation itself but in how we derive and interpret the values for each factor. The first two or three terms now have empirically established ranges thanks to advances in astronomy. We know approximately how many stars form each year in our galaxy and, with the discovery of thousands of exoplanets, we have reasonable estimates for how many stars have planetary systems.

But beyond these initial terms, we enter the realm of speculation. We have a sample size of exactly one—Earth—for estimating the likelihood that life emerges on habitable planets, that such life evolves intelligence, that intelligence produces technology, and that technology creates detectable signals. Extrapolating from a single data point to a universal probability distribution is not merely uncertain; it is methodologically unsound.

Consider the factor fl—the fraction of habitable planets that develop life. On Earth, life emerged relatively quickly after the planet cooled enough to support liquid water, suggesting that abiogenesis might be common under Earth-like conditions. But this is an inference from a single case. It's equally plausible that life's emergence on Earth was a freak accident, a one-in-a-trillion event that happened to occur on our planet. Without other examples, we cannot distinguish between these possibilities.

The same problem applies to each subsequent factor. The evolution of human-like intelligence on Earth required a specific sequence of contingent events—the Cambrian explosion, the K-T extinction that removed dominant reptiles, the particular social dynamics of primate evolution—that might be vanishingly rare on a cosmic scale. Or, conversely, intelligence like ours might emerge commonly but take forms we wouldn't recognize. Without comparative data, we cannot know.

This statistical overreach becomes even more apparent when we consider how Drake Equation calculations are typically presented. Scientists and science communicators often offer a range of values for each factor, calculate the resulting range for N, and then treat the midpoint or average of this range as a reasonable estimate. But this approach subtly transforms uncertainty into pseudoprecision. The wide range—which might span from "effectively zero" to "billions of civilizations"—gets reduced to a single most likely value, creating an illusion of knowledge where none exists.

The mathematician John Allen Paulos calls this the "risk of false mathematical precision"—using numbers to create an impression of objectivity that masks underlying assumptions and uncertainties. When applied to extraterrestrial intelligence, this false precision can lead us to treat the "paradox" of alien absence as a genuine logical problem requiring explanation, rather than recognizing it as an artifact of our statistical methodology.

Sample Size: The Problem of N=1

THE FUNDAMENTAL STATISTICAL flaw in our approach to extraterrestrial intelligence can be summarized simply: we are attempting to derive universal principles from a sample size of one. Everything we know about life, intelligence, technology, and civilization comes from Earth's single example. Our disciplines of biology, psychology, sociology, and history are essentially the study of one planetary case study.

Statisticians have a term for the problem of limited samples: sampling bias. When we select data points that aren't representative of the whole population, we draw distorted conclusions. Earth represents the ultimate sampling bias—we know about life here precisely because we evolved here. We have no access to the vast majority of potential data points: planets where life never emerged, where it emerged but remained microbial, where intelligence evolved but didn't create technology, or where radically different forms of organization emerged that we wouldn't recognize as life or intelligence at all.

This problem becomes clear when we examine how variables interact in complex systems. In statistical analysis, variables that appear independent may actually be deeply correlated in ways that only become apparent with large sample sizes. For example, astrobiology often treats the emergence of life, the evolution of intelligence, and the development of technology as independent probabilities that can be multiplied together. But what if these factors are correlated—either positively (planets suitable for life are also suitable for intelligence) or negatively (most life-bearing planets develop ecological stability that prevents the emergence of technology-creating species like humans)?

Without comparative examples, we cannot know whether Earth represents a typical pathway for life and intelligence or an extreme outlier. We fall victim to what statisticians call the "streetlight effect"—looking for answers only where our existing knowledge illuminates the search, not necessarily where answers might be found.

The N=1 problem extends beyond the question of how common alien civilizations might be to the question of how we might recognize them. Our instruments and search strategies are calibrated to detect signatures similar to human civilization: radio emissions, atmospheric modifications, energy utilization patterns that resemble what we might do if we were more advanced. But this approach assumes that all technological civilizations follow similar developmental trajectories and make similar choices about energy and communication—an assumption that lacks empirical support even within Earth's single example of technological civilization, where multiple pathways of development have occurred in different societies.

Some astrobiologists have attempted to address the N=1 problem by studying "natural experiments" on Earth—cases where life evolved independently in isolated environments or where intelligence evolved convergently in different lineages (such as corvids, cetaceans, cephalopods, and primates). These studies provide valuable insights into evolutionary constraints and possibilities, but they still operate within Earth's specific conditions. They cannot tell us about possibilities that never had the chance to emerge in Earth's environment.

The statistical challenge isn't merely that we lack data—it's that we may lack imagination about what data would even be relevant. Our conception of the probability space may be far too constrained by our terrestrial experience, causing us to miss vast regions of possibility simply because they don't fit our conceptual frameworks.

From Probability to Certainty: The Rhetoric of Alien Abundance

DESPITE THESE FUNDAMENTAL statistical limitations, discussions of extraterrestrial intelligence often slide from careful statements of possibility to confident assertions of probability or even certainty. Consider these statements from respected scientists and science communicators:

"There are something like a few hundred billion stars in the Milky Way galaxy, and if one in five has a planet, and if one in a hundred planets is habitable, and one in a hundred habitable planets eventually develops life, and one in a hundred planets with life eventually develops intelligent life, then there are still millions of planets with intelligent life in the galaxy." — Carl Sagan

"To my mathematical brain, the numbers alone make thinking about aliens perfectly rational." — Stephen Hawking

"It is not a question of whether there is intelligent life elsewhere in the universe, but when we will find it." — Seth Shostak, SETI Institute

These statements exemplify what I call the "rhetoric of alien abundance"—a discursive pattern that transforms the mere possibility of extraterrestrial intelligence into an overwhelming probability through a series of linguistic and mathematical moves. This rhetoric typically follows a predictable structure:

1. Begin with the vastness of the cosmos (billions of galaxies, each with billions of stars)
2. Apply seemingly conservative probabilities to each step from star to intelligence
3. Demonstrate that even with these conservative estimates, the numbers remain enormous
4. Conclude that intelligence must be common, making our apparent isolation paradoxical

This rhetorical pattern is so common in popular science that it has become almost ritualistic—a required preamble to any discussion of the search for extraterrestrial intelligence. Its persuasive power comes from the way it combines genuine astronomical facts (the number of stars) with speculative probabilities, treating both as equally established.

The rhetoric of alien abundance serves important social and psychological functions. It situates humanity within a cosmic context, suggesting that we are not alone in the universe. It validates the search for extraterrestrial intelligence as a rational scientific endeavor rather than a fringe pursuit. And it offers hope that we might someday join a community of intelligent species, transcending our isolation.

But these functions do not make the rhetoric scientifically sound. The slide from possibility to probability to virtual certainty represents a form of what philosophers call the "conjunction fallacy"—the mistaken belief that specific scenarios are more probable than general ones. Each additional condition we place on alien intelligence (that it uses radio communication, that it values expansion, that it exists contemporaneously with us) makes the overall scenario less probable, not more. Yet our intuition often suggests the opposite—that the detailed scenario we've constructed must be common given the vastness of space.

The rhetoric of alien abundance also exhibits a peculiar doublethink regarding anthropocentrism. On one hand, it uses Earth as the template for estimating every probability in the calculation. On the other hand, it presents the resulting high probability as a rejection of human specialness or cosmic importance. This contradiction—using Earth as the measure of all things while claiming to transcend Earth-centered thinking—reveals how deeply our extraterrestrialist framework is rooted in anthropocentric assumptions even as it claims to move beyond them.

The Great Filter: Catastrophe as Explanation

AMONG THE MOST INFLUENTIAL solutions to Fermi's Paradox is Robin Hanson's "Great Filter" hypothesis. According to this idea, the development of advanced, spacefaring civilizations must be extremely rare because there exists some developmental step—a "filter"—that is extraordinarily difficult to pass. The unsettling implication is that this filter might lie ahead of us rather than behind us; perhaps civilizations typically destroy themselves once they reach our current technological stage.

The Great Filter hypothesis has captured both scientific and popular imagination, spawning research programs, science fiction narratives, and philosophical discussions about humanity's prospects. Its appeal lies partly in its explanatory power—it offers a clean solution to the apparent paradox—and partly in its dramatic implications for human civilization. It transforms the search for extraterrestrial intelligence into an existential investigation: by understanding why we don't observe aliens, we might learn crucial information about our own survival prospects.

However, the Great Filter concept exemplifies how proposed solutions to Fermi's Paradox often smuggle in anthropocentric assumptions. The hypothesis presupposes a universal developmental trajectory for intelligent life—a sequence of steps from simple cells to complex organisms to technological civilization to interstellar expansion. This trajectory is derived entirely from Earth's history and human civilizational development, yet is projected as a universal pathway.

Consider the assumption that technological civilizations will naturally expand beyond their home planets if they can. This expectation reflects specific aspects of human history—particularly European colonialism and the American frontier experience—rather than a universal characteristic of intelligence. Even within human history, not all societies have prioritized territorial expansion; some have chosen stability, sustainability, or cultural refinement over growth. There is no reason to assume that alien intelligences would share the expansionist imperative that characterizes certain human civilizations.

Similarly, the conception of a filter as a single catastrophic event or developmental hurdle reflects human cognitive biases. We tend to look for dramatic, singular causes rather than complex, multi-factor explanations. This narrative preference shapes how we imagine extraterrestrial development—as a hero's journey with a crucial test to pass rather than as a complex, contingent process with multiple pathways and possibilities.

The Great Filter hypothesis also reveals the subtle ways in which Fermi's Paradox functions as a mirror for contemporary anxieties. The post-World War II origins of the paradox reflected Cold War concerns about nuclear annihilation. Today's discussions often focus on climate change, artificial intelligence, or biotechnology as potential filters that might explain alien absence and threaten human survival. The aliens are absent for precisely the reasons we fear we might not persist—a remarkable coincidence that suggests we are projecting our anxieties onto the cosmic screen.

This is not to dismiss the genuine existential risks that advanced technologies pose. These concerns deserve serious attention. But we should recognize how our attempt to solve Fermi's Paradox often serves as a displacement activity—a way to address our fears about human civilization's prospects through the safer, more abstract question of alien absence.

The Zoo Hypothesis: Cosmic Conservation

or Cosmic Theology?

ANOTHER POPULAR SOLUTION to Fermi's Paradox is the "Zoo Hypothesis," proposed by John Ball in 1973. According to this idea, extraterrestrial civilizations exist but have chosen not to contact or interfere with Earth, preserving us in something like a cosmic wildlife reserve or anthropological research site. Various elaborations suggest that we might be subject to a "prime directive" (borrowing from Star Trek) that prohibits contact with less developed civilizations, or that Earth might be a controlled experiment or reality show for alien entertainment.

The Zoo Hypothesis appears to offer a respectful solution to the paradox—one that acknowledges both the possibility of advanced civilizations and the lack of observed contact while avoiding doomsday scenarios. It has the additional virtue of being unfalsifiable; the better aliens are at hiding, the less evidence we would have of their existence.

However, this hypothesis reveals perhaps the most profound anthropocentrism embedded in our extraterrestrialist thinking. It essentially recasts the Judeo-Christian concept of a hidden God watching over humanity into secular, scientific terms. The aliens of the Zoo Hypothesis serve the same psychological function as the divine observer—they witness our struggles, understand our true nature, and exercise power over our fate while remaining invisible to us. They provide a cosmic audience for human affairs, ensuring that our activities have significance beyond our isolated planet.

The parallels with religious cosmology extend further. The hypothesis often includes the possibility that aliens might reveal themselves when humanity reaches some developmental milestone—perhaps global peace, technological singularity, or interstellar capability. This "cosmic coming of age" narrative mirrors religious concepts of revelation conditional upon moral or spiritual advancement. Even the language used to discuss the hypothesis—references to "watchers," "guardians," or "observers"—echoes religious terminology.

This theological subtext suggests that the Zoo Hypothesis functions less as a scientific explanation than as a secular mythology addressing existential concerns about cosmic meaning and human significance. It provides conceptual continuity between traditional religious frameworks and scientific cosmology, preserving the comforting notion that we are seen and known by greater intelligences even as it replaces supernatural entities with technological ones.

Beyond its theological resonances, the Zoo Hypothesis exhibits what might be called "conservation chauvinism"—the projection of contemporary Western environmental and anthropological ethics onto alien civilizations. The very concept of a "zoo" or "nature reserve" reflects recent human attempts to mitigate our destructive impact on other species and cultures. It assumes that aliens would share our belated concern for non-interference and preservation, rather than the exploitative attitudes that have characterized much of human history.

It also centers humanity as worthy of special protection or observation—an assumption that seems dubious given our brief existence and limited technological development. Why would aliens go to extraordinary lengths to hide from us specifically, rather than from the millions of other species that have existed on Earth over its history? The Zoo Hypothesis tacitly assumes human exceptionalism even as it places humans in a subordinate position relative to hypothetical alien observers.

Perhaps most tellingly, the hypothesis projects contemporary academic and conservation institutions—zoos, wildlife reserves, anthropological field sites—onto a cosmic scale. It imagines aliens essentially as cosmic scholars or conservationists, sharing the values and methodologies of modern Western scientific and educational establishments. This institutional projection reveals how thoroughly our extraterrestrialist imagination is shaped by our current social structures and values.

The Simulation Hypothesis: Metaphysics Masquerading as Science

A MORE RECENT ADDITION to Fermi Paradox solutions is the Simulation Hypothesis, popularized by philosopher Nick Bostrom's 2003 paper "Are You Living in a Computer Simulation?" According to this hypothesis, what we experience as reality is actually a computer simulation created by an advanced civilization. The apparent absence of aliens in our universe is simply a feature of the simulation—either aliens weren't programmed into it, or they exist outside the simulation in the "real" universe.

This hypothesis has gained prominence in both academic philosophy and popular culture, with figures like Elon Musk suggesting that the probability we live in "base reality" is one in billions. Its appeal lies partly in its compatibility with both scientific and religious worldviews—it can be interpreted as either a technological version of divine creation or as a purely materialist explanation requiring no supernatural elements.

The Simulation Hypothesis represents a fascinating case of what philosopher Karl Popper called "pseudoscience"—a theory that has the appearance of scientific reasoning but lacks empirical falsifiability. It makes no testable predictions that would distinguish our experience in a simulation from experience in "base reality." Any apparent evidence against the simulation could simply be explained as a feature programmed into the simulation itself.

This unfalsifiability reveals the hypothesis's true nature: it is metaphysics presented in the language of computer science. Like traditional metaphysical systems, it attempts to explain the fundamental nature of reality through reason rather than empirical investigation. The digital computers and simulations it invokes serve the same function in contemporary thought that divine architecture served in earlier cosmologies—they provide a conceptually familiar framework for explaining the universe's fundamental structure.

The simulation hypothesis exemplifies what philosopher Alfred North Whitehead called the "fallacy of misplaced concreteness"—treating abstract concepts as if they were concrete realities. It takes the human-created concept of computer simulation, which exists within our reality, and projects it as the explanation for reality itself. This circular reasoning uses the part to explain the whole, the created to explain the creator.

From the perspective of extraterrestrialism, the simulation hypothesis represents perhaps the ultimate anthropocentric projection. It suggests that the entire universe operates according to principles analogous to human-created technology. It imagines post-humans or aliens whose motivations for creating simulations mirror our own reasons for creating video games or scientific models. And it centers human experience as the apparent focus of the simulation, since we are the conscious entities experiencing it from within.

Moreover, the hypothesis reveals the subtle influence of contemporary technological metaphors on our cosmic thinking. Earlier generations conceived of the universe as a great machine or clockwork; today we conceive of it as a computer simulation. Each era projects its dominant technological paradigm onto reality itself, revealing the contingency of these metaphors rather than their objective truth.

The simulation hypothesis offers a "solution" to Fermi's Paradox by essentially dissolving the paradox altogether—declaring that our observations of cosmic emptiness are not observations of the actual universe at all, but merely features of a programmed environment. This move highlights the extreme measures to which we will go to preserve the belief in abundant alien intelligence when confronted with contrary evidence. Rather than questioning our initial probability estimates, we question reality itself.

The Transcension Hypothesis: Digital Rapture

AMONG THE MORE SOPHISTICATED responses to Fermi's Paradox is the Transcension Hypothesis, developed by futurist John Smart. This hypothesis suggests that advanced civilizations don't expand outward into space but rather inward, toward states of increasingly dense energy use, computation, and complexity. Eventually, these civilizations approach black-hole-like conditions where they can conduct vastly more computation with less energy, potentially even linking to other universes or dimensions. According to this hypothesis, the reason we don't observe alien megastructures or interstellar expansion is that such activities represent primitive, inefficient approaches that advanced civilizations quickly abandon in favor of "transcension" to higher planes of existence.

The Transcension Hypothesis has gained traction among futurists and some scientists because it extrapolates from observable technological trends—miniaturization, virtualization, increasing computational efficiency—rather than relying solely on speculative alien psychology. It also aligns with physical principles such as the tendency of complex adaptive systems to maximize computational capacity while minimizing energy expenditure.

Yet for all its apparent scientific sophistication, the Transcension Hypothesis reveals distinctly human philosophical and religious underpinnings. Its vision of civilizations "ascending" to higher dimensional states bears striking resemblance to religious concepts of transcendence, particularly the Christian rapture or Eastern notions of liberation from material existence. The hypothesis essentially secularizes spiritual transcendence, replacing divine grace with computational physics while preserving the narrative of escape from material limitations to a higher state of being.

This spiritual subtext becomes clearer when we examine the language used to describe "transcended" civilizations. They are often characterized as achieving "godlike" computational capacity, "omniscience" within their light cones, or "immortality" through substrate independence. The hypothesis transforms traditional theological attributes into technological achievements, preserving the structure of religious thought while updating its content for a scientific age.

The hypothesis also projects specifically contemporary technological trends onto alien development. It assumes that the miniaturization and virtualization characterizing recent human computing will continue indefinitely and represents a universal developmental trajectory. This projection ignores the contingency of our technological history and the possibility that different starting conditions, resources, or evolutionary pathways might lead to entirely different technological trajectories.

Perhaps most tellingly, the Transcension Hypothesis preserves what might be called the "adventure narrative" of human technological development—the sense that we are on a journey toward ever-greater knowledge, power, and transcendence of natural limitations. This narrative reflects Western cultural values of progress, mastery, and individual liberation that may not be shared by all human cultures, let alone by alien species with different evolutionary and cultural histories.

Like other proposed solutions to Fermi's Paradox, the Transcension Hypothesis functions as a mirror for human aspirations rather than as an objective assessment of cosmic possibilities. It addresses our hope for cosmic company by suggesting that advanced aliens exist but have transformed themselves into forms we cannot yet perceive. It addresses our fear of technological stagnation by suggesting that our current development trajectory leads toward transcendent capacities. And it addresses our existential anxieties by suggesting that intelligence can escape the material constraints that currently limit human experience.

These functions are not inherently problematic—they provide conceptual resources for imagining positive technological futures. But we should recognize them as projections of human concerns rather than as revelations of universal developmental principles. The transcension narrative may tell us more about our hopes for our own civilization than about the actual nature of extraterrestrial intelligence.

Beyond Fermi: Rethinking the Question

THE PROPOSED SOLUTIONS to Fermi's Paradox—from the Great Filter to the Zoo Hypothesis to the Simulation Hypothesis to the Transcension Hypothesis—all attempt to reconcile the perceived high probability of alien existence with the observed lack of contact. They accept the basic framing of the paradox and offer various ways to resolve the apparent contradiction. But what if the paradox itself rests on faulty premises? What if the question itself needs rethinking?

The structure of Fermi's Paradox assumes:

1. That the emergence of intelligence like ours is relatively common
2. That technological development follows a fairly predictable trajectory
3. That expansion and communication are natural behaviors for advanced intelligence
4. That we would recognize evidence of alien civilization if it existed

Each of these assumptions deserves critical examination.

The first assumption—that intelligence like ours commonly emerges—projects terrestrial evolutionary history onto the cosmos. It ignores the possibility that human-like intelligence might be an evolutionary contingency rather than an inevitability, the result of specific environmental pressures and historical accidents that may be vanishingly rare on a cosmic scale. The fact that complex intelligence evolved only once on Earth, despite billions of years of evolution and millions of species, should give us pause before assuming it represents a common outcome.

The second assumption—that technological development follows a predictable trajectory—ignores the diversity of human technological histories and the role of cultural and environmental factors in shaping technological choices. Different human societies have prioritized different technologies based on their values, resources, and needs. Some have deliberately rejected technologies that others embraced. Even within industrialized societies, technological determinism—the idea that technology develops along a single, inevitable path—has been thoroughly critiqued by historians and sociologists of technology. If technological development isn't deterministic within human history, why would it be across different species and planets?

The third assumption—that expansion and communication are natural behaviors for intelligence—projects specific human tendencies onto alien psychology. Not all human societies have valued expansion; some have prioritized stability, sustainability, or cultural refinement. Communication across differences presents enormous challenges even between human cultures; the barriers to interspecies communication would be far greater. We cannot assume that all forms of intelligence would share our particular tendencies toward growth and outreach.

The fourth assumption—that we would recognize evidence of alien civilization—may be the most questionable of all. Our search methods are calibrated to detect technologies similar to our own, communications in frequency ranges we consider logical, and engineering projects that follow principles we understand. But alien technologies, communications, and engineering might operate according to principles entirely different from those we've discovered. As Arthur C. Clarke famously observed, any sufficiently advanced technology would be indistinguishable from magic to us—and we might not even recognize it as technology at all.

Rethinking these assumptions transforms Fermi's "Where is everybody?" from a paradox requiring solution to a question highlighting the limits of our knowledge and imagination. The apparent absence of evidence for alien civilization may tell us more about the constraints of our search methods and conceptual frameworks than about the actual distribution of intelligence in the cosmos.

A more productive approach might be to embrace what philosopher of science Nancy Cartwright calls "the dappled world"—the recognition that our scientific theories and models apply well to some domains and poorly to others, that reality is not fully captured by any single framework or methodology. Applied to extraterrestrial intelligence, this perspective suggests that we should develop multiple, complementary approaches to the question, each with explicit recognition of its assumptions and limitations.

Such approaches might include:

1. **Multiple origin scenarios** that don't assume life elsewhere follows Earth's chemical or evolutionary pathways
2. **Non-technological intelligence models** that consider forms of organization and adaptation not centered on tool use or environmental manipulation
3. **Non-expansion hypotheses** that explore how intelligence might flourish without prioritizing growth or dispersion
4. **Alternative communication frameworks** that consider fundamentally different modes of information exchange
5. **Non-detection explanations** that consider the limitations of our observational capacities and interpretive frameworks

These approaches don't answer Fermi's question definitively, but they transform it from a paradox demanding resolution to an open inquiry revealing the boundaries of our knowledge. They replace the false certainty of probability calculations with the generative uncertainty of multiple possibilities, each offering different insights into the nature of intelligence and its place in the cosmos.

The Value of Not Knowing

PERHAPS THE MOST PROFOUND lesson of Fermi's Paradox is not any particular solution but the recognition of how much we don't—and perhaps cannot—know about cosmic intelligence. This recognition is not a failure but an opportunity—a chance to develop what philosopher Rebecca Newberger Goldstein calls "the maturity of mind to tolerate, even to welcome, the uncertainty that is a corollary of our finitude."

The search for extraterrestrial intelligence represents one of humanity's noblest endeavors—an attempt to transcend our planetary isolation and connect with possible cosmic companions. This search should continue, informed by the best science and the most imaginative speculation. But it should proceed with awareness of its limitations and biases, with openness to possibilities beyond our current conception, and with recognition that the absence of certainty is not a problem to be solved but a condition to be embraced.

In our eagerness to answer the question "Are we alone?", we have sometimes confused probability with certainty, projection with discovery, and human tendencies with universal principles. By recognizing these confusions, we can develop a more nuanced, more humble, and ultimately more accurate approach to the question of extraterrestrial intelligence—one that acknowledges the vast cosmos of possibility while respecting the boundaries of our knowledge.

Fermi's question—"Where is everybody?"—may not have a single correct answer. It may have many answers, or none that we can comprehend with our current conceptual tools. But in asking it, and in exploring the many possible responses, we deepen our understanding not just of potential cosmic companions but of ourselves—our assumptions, our limitations, and our place in a universe that continues to exceed our grasp even as it inspires our curiosity.

The true value of Fermi's Paradox may lie not in solving it but in living with it—in allowing its apparent contradiction to expand our imagination, challenge our assumptions, and remind us of the vastness that lies beyond our current understanding. In embracing this uncertainty, we may come closer to the cosmic perspective that has always been the deeper purpose of our gaze toward the stars.

Chapter 4: Of Dyson Spheres and Kardashev Civilizations: The Folly of Ignoring Human Geography

"Any sufficiently advanced technology is indistinguishable from magic." — Arthur C. Clarke

"Any sufficiently advanced extraterrestrial intelligence is indistinguishable from the laws of physics." — Proposed corollary by Karl Schroeder

Engineering the Cosmos: The Megastructure Imagination

In 1960, physicist Freeman Dyson published a brief paper titled "Search for Artificial Stellar Sources of Infrared Radiation" in the journal Science. In this modest four-page article, Dyson proposed that advanced technological civilizations would inevitably face energy constraints and would likely solve this problem by constructing enormous structures around their stars to capture most or all of the stellar energy output. These hypothetical "Dyson spheres" would absorb visible light while reradiating waste heat as infrared radiation, potentially making them detectable from Earth through astronomical observation.

Dyson's speculation marked a turning point in how scientists conceptualized the search for extraterrestrial intelligence. Rather than focusing solely on intentional communications like radio signals, researchers could search for the unintentional technosignatures of cosmic engineering—the physical footprints of advanced civilizations reshaping their environments on stellar scales. This approach seemed more reliable; while aliens might choose not to communicate with us, they presumably couldn't hide the infrared signatures of their energy-harvesting megastructures.

Around the same time, Soviet astronomer Nikolai Kardashev proposed a three-tiered classification system for technological civilizations based on their energy utilization:

- **Type I civilizations** harness all available energy on their home planet (approximately 10^{16} watts)
- **Type II civilizations** harness all energy from their host star (approximately 10^{26} watts)
- **Type III civilizations** harness all energy from their entire galaxy (approximately 10^{36} watts)

This "Kardashev Scale" has since become a standard framework for conceptualizing the potential trajectory of technological advancement, both human and alien. It has been expanded by subsequent thinkers to include hypothetical Type IV civilizations (controlling multiple galaxies or galaxy clusters) and Type V civilizations (controlling the entire observable universe).

Together, Dyson spheres and the Kardashev Scale have shaped scientific and popular imagination about advanced alien civilizations for over half a century. They have inspired countless science fiction narratives, from Larry Niven's "Ringworld" to Iain M. Banks' "Culture" series. They have guided astronomical searches for unusual infrared signatures around stars. And they have provided a conceptual roadmap for humanity's own potential long-term development.

But these frameworks, powerful as they are, contain hidden assumptions about the nature of technological progress, the values of advanced intelligences, and the universal desirability of certain developmental trajectories. These assumptions tell us more about mid-20th century industrial capitalism than about the likely nature of extraterrestrial civilizations. By examining them critically, we can begin to appreciate how our extraterrestrialist imagination has been constrained by historically and culturally specific conceptions of development and progress.

The Industrial Capitalism of the Stars

THE KARDASHEV SCALE, despite its cosmic scope, is remarkably provincial in its conception of advancement. It measures civilizational progress along a single axis: energy consumption. This metric reflects the specific historical context in which Kardashev developed his scale—the industrial rivalry between the Soviet Union and the United States during the Cold War, when both superpowers equated national power with industrial output and resource utilization.

This energy-centric measure of advancement was not arbitrary for its time. Both Soviet communism and American capitalism, despite their ideological differences, shared a fundamental commitment to industrial production, resource extraction, and economic growth as markers of civilizational success. Higher energy consumption correlated with greater military capability, economic output, and geopolitical influence. The Space Race itself was as much about demonstrating industrial-technological supremacy as about scientific exploration.

Projecting this industrial paradigm onto extraterrestrial civilizations, however, reveals a profound anthropocentrism in our cosmic imagination. It assumes that all technological species, regardless of their biology, evolutionary history, planetary conditions, or cultural values, will follow a developmental trajectory similar to mid-20th century humanity—prioritizing ever-increasing energy consumption and physical engineering capabilities.

Consider how the Kardashev Scale implicitly privileges certain forms of technology and organization:

1. **Physical over informational technologies**: The scale emphasizes energy-intensive physical construction and resource manipulation rather than efficient information processing or social technologies.
2. **Quantitative over qualitative development**: Progress is measured by how much energy is used, not how wisely it is utilized or for what purposes.
3. **Centralized over distributed systems**: The scale assumes civilizations that act as unified agents capable of coordinating vast engineering projects, rather than diverse, decentralized networks with multiple competing or complementary approaches.
4. **Expansion over intensification**: Development is equated with physical growth and expansion rather than internal refinement or efficiency improvements.

These preferences aren't universal truths about technological development but reflections of particular industrial capitalist values prevalent when the scale was conceived. They reveal more about the political economy of Cold War superpowers than about the likely trajectory of intelligence across the cosmos.

Dyson spheres embody similar assumptions. The concept presumes that advanced civilizations will:

1. **Prioritize energy capture** over other developmental possibilities
2. **Value growth** over stability or sustainability
3. **Prefer centralized mega-engineering** over distributed, smaller-scale solutions
4. **Maintain human-like resource consumption patterns** even as they advance technologically

These presumptions project a specifically industrial capitalist relationship to resources onto the cosmos—one characterized by extraction, exploitation, and expansion rather than equilibrium, efficiency, or ecological integration.

It's worth noting that even within human civilization, these industrial values have not been universal across cultures or historical periods. Many societies have prioritized stability over growth, spiritual development over material accumulation, or ecological harmony over resource exploitation. The industrial capitalist paradigm represents one cultural approach among many, not a universal or inevitable stage of development.

By encoding this specific paradigm into our frameworks for conceptualizing advanced civilizations, we constrain our ability to recognize or imagine alternatives. We search for alien intelligence that resembles our industrial ambitions writ large rather than opening ourselves to the possibility of fundamentally different approaches to technological development and resource relationship.

The Assumption of Perpetual Growth

PERHAPS THE MOST DUBIOUS assumption underlying both the Kardashev Scale and Dyson sphere concepts is the presumption of perpetual growth as both desirable and possible. These frameworks envision civilizations that continuously expand their energy consumption from planetary to stellar to galactic scales, with no inherent limits or alternative pathways.

This assumption mirrors the growth imperative of industrial capitalism, which treats economic expansion as both a necessity and a virtue. Modern industrial economies are structured around the expectation of continuous GDP growth, increasing production and consumption, and expanding resource utilization. Stagnation or contraction are viewed as failures rather than potential alternative states.

Yet perpetual physical growth faces fundamental limits in any finite system. Earth's ongoing ecological crisis demonstrates the tensions between continuous industrial expansion and planetary boundaries. Resource depletion, pollution, climate disruption, and biodiversity loss all point to the incompatibility of unlimited growth with the finite capacity of ecological systems.

Proponents of cosmic megastructures might argue that precisely because of these planetary limitations, advanced civilizations must expand beyond their home worlds to maintain growth. This argument recasts cosmic engineering as the inevitable solution to growth constraints rather than questioning the growth imperative itself. It transforms space into a new frontier for resource exploitation rather than an opportunity to reimagine civilizational purpose and structure.

This narrative has troubling parallels with colonial expansion during the age of European imperialism. When resource limitations, social contradictions, or environmental degradation constrained growth in core regions, the solution was geographic expansion into "new worlds" rather than fundamental restructuring of economic systems or cultural values. The cosmic frontier serves a similar conceptual function in our extraterrestrialist imagination—a safety valve for growth limitations rather than a catalyst for systemic reimagining.

The perpetual growth assumption also ignores the possibility that technological advancement might fundamentally change resource requirements rather than simply scaling them up. Consider how digital technologies have dematerialized many human activities, replacing physical goods and services with information-based alternatives. Advanced civilizations might develop technologies that reduce rather than increase their material and energy footprints, achieving more sophisticated capabilities with fewer resources.

Even more fundamentally, the growth assumption projects a particular conception of desire and satisfaction onto alien psychology. It presumes that advanced intelligences would always want more—more energy, more matter, more physical infrastructure—rather than reaching states of equilibrium or sufficiency. This presumption reveals how deeply consumer capitalism has shaped our conception of motivation and progress, even in supposedly objective scientific frameworks.

Alternatives to the Kardashev Trajectory

WHAT MIGHT ADVANCED intelligence look like if it doesn't follow the Kardashev trajectory of ever-increasing energy consumption and cosmic engineering? What alternative developmental pathways might extraterrestrial civilizations pursue? These questions require us to step outside the industrial capitalist paradigm and imagine different relationships between intelligence, technology, and environment.

Post-Scarcity Without Mega-Engineering

ONE POSSIBILITY IS that technological advancement might lead to post-scarcity economics without requiring stellar-scale engineering. Advanced nanotechnology, for instance, might enable the creation of complex goods with minimal resource inputs. Perfect recycling systems might eliminate waste and create closed material loops. Fusion energy or advanced solar capture might provide abundant energy without Dyson spheres.

In such scenarios, civilizations might achieve material abundance within planetary boundaries, making stellar-scale resource capture unnecessary. Their technological development would focus on efficiency, circularity, and dematerialization rather than gross energy consumption. They might appear less advanced on the Kardashev Scale while actually possessing more sophisticated technologies and higher quality of life.

The absence of detectable megastructures around stars might not indicate the absence of advanced civilizations but rather the presence of civilizations advanced enough to transcend the need for cosmic-scale resource exploitation. Our search for Dyson spheres might be akin to 18th-century observers looking for massive coal stockpiles as evidence of development, completely missing the possibility of nuclear energy.

Inward-Turning Development

ANOTHER ALTERNATIVE is what futurist John Smart calls "transcension"—the possibility that advanced civilizations turn inward rather than outward, developing technologies that operate at increasingly microscopic scales with increasingly efficient energy use. Rather than expanding physically into space, such civilizations might explore inner space—quantum realms, simulated realities, or modes of experience that don't require extensive physical infrastructure.

This pathway suggests that the most advanced technologies might be the least detectable from afar. A civilization processing vast amounts of information in quantum computing systems might have a tiny physical footprint compared to one building megastructures, yet it could be far more advanced in terms of knowledge, capabilities, and experiences.

From an anthropocentric perspective focused on physical engineering, such inward-turning civilizations might appear undeveloped or stagnant. From their own perspective, they might be exploring frontiers far more interesting and valuable than outer space. Our search methods, calibrated to detect large-scale manipulation of matter and energy, would miss these sophisticated but physically unobtrusive intelligences.

Non-Technological Forms of Advanced Intelligence

PERHAPS THE MOST RADICAL alternative is that some forms of intelligence might advance along entirely non-technological pathways. Our equation of advancement with technological sophistication reflects a specifically human evolutionary strategy—the use of external tools to modify environments. But different evolutionary histories might produce intelligences that develop other capabilities instead.

Consider possibilities like:

1. **Biological rather than technological evolution**: Species that continue to evolve biologically rather than supplementing their capabilities with technology
2. **Collective intelligence systems**: Organisms that develop

sophisticated neural networks or distributed cognition without external technology

3. **Ecological integration rather than control**: Intelligences that advance by developing more complex symbiotic relationships with their environments rather than by dominating them
4. **Perception and consciousness enhancement**: Beings that evolve expanded awareness or novel forms of consciousness rather than external tools

These pathways might produce intelligences with capabilities we would consider "advanced" but would be undetectable through technosignature searches. A species with profound consciousness or perception beyond human experience might appear technologically primitive while possessing forms of knowledge and awareness we cannot comprehend.

Our difficulty imagining such alternatives reveals the constraints of human evolutionary experience on our conception of intelligence. Having evolved as tool-using primates, we naturally associate intelligence with technological manipulation. But this association may be contingent rather than universal—one possible pathway among many rather than the inevitable trajectory of all advanced minds.

Earth's Civilizational Diversity as Caution

WE NEED NOT SPECULATE about completely alien scenarios to recognize the limitations of the Kardashev framework. Earth's own civilizational diversity offers ample evidence that technological development follows multiple trajectories rather than a single, linear path.

Consider the diverse ways human societies have developed and applied technologies:

1. **China's historical trajectory** included sophisticated technologies like printing, gunpowder, and complex hydraulic engineering systems centuries before Europe, yet it did not develop industrial capitalism until recently. This wasn't due to stagnation but reflected different priorities and social structures.
2. **Indigenous societies in the Americas** developed complex agricultural systems, urban centers, and sophisticated ecological knowledge without many technologies Europeans considered essential, such as metallurgy or wheeled transport. These civilizations optimized for different values and environmental relationships.
3. **Japan's Tokugawa period** saw deliberate technological simplification in some areas (including the abandonment of firearms) while advancing in others, demonstrating that technological development can be selective rather than comprehensive.
4. **Contemporary Amish communities** carefully evaluate technologies based on their impacts on community and religious values, adopting some modern innovations while rejecting others that would undermine their social fabric.

These examples demonstrate that technological development is not a single, determined pathway but a series of choices reflecting cultural values, environmental contexts, and social priorities. Different societies prioritize different capabilities and make different trade-offs between technological possibilities.

Extrapolating from this terrestrial diversity, we might expect extraterrestrial civilizations to exhibit even greater variation in their developmental trajectories. Aliens would likely make technological choices based on their specific biology, environment, cognitive structures, and cultural values—all potentially radically different from human parameters.

Some might prioritize technologies that enhance collective consciousness while neglecting physical engineering. Others might develop sophisticated environmental management systems while having little interest in space travel. Still others might focus on miniaturization and efficiency rather than expansion and resource accumulation.

The Kardashev Scale, by reducing this potential diversity to a single metric of energy consumption, impoverishes our conception of how intelligence might manifest across the cosmos. It privileges one particular developmental trajectory—that of expanding industrial resource exploitation—over countless other possibilities.

Indigenous and Non-Western Perspectives

THE LIMITATIONS OF the Kardashev-Dyson framework become even clearer when we consider indigenous and non-Western perspectives on technology, progress, and human-environment relationships. These diverse traditions offer alternative conceptual frameworks that might better capture some possible extraterrestrial approaches to development.

Many indigenous cosmologies emphasize:

1. **Relational rather than extractive resource approaches**: Viewing natural resources as entities with which one maintains reciprocal relationships rather than inert materials to be exploited
2. **Circular rather than linear conceptions of time and**

development: Prioritizing cycles and renewal over constant progress toward new states

3. **Distributed knowledge systems**: Valuing collective, intergenerational knowledge transmission over centralized data accumulation
4. **Place-based rather than universal technologies**: Developing highly adapted solutions for specific environments rather than standardized approaches applied universally

These frameworks suggest radically different trajectories for advanced civilizations than those implied by the Kardashev Scale. An advanced civilization operating according to relational resource ethics might develop sophisticated technologies for maintaining planetary equilibrium rather than extracting ever-greater amounts of energy. A society with circular time conceptions might focus on refinement of existing systems rather than continuous expansion into new frontiers.

Similarly, non-Western philosophical traditions offer alternative conceptions of progress and advancement:

1. **Buddhist perspectives** prioritize internal development of consciousness over external technological control
2. **Daoist approaches** emphasize harmony with natural systems rather than domination of them
3. **Various African philosophies** center communal well-being and social technologies over individual or centralized technological power

These traditions aren't simply "less advanced" versions of Western technological development but fundamentally different approaches to the relationship between intelligence, technology, and environment. They suggest that advancement might be measured in forms of consciousness, social harmony, or ecological integration rather than energy consumption or engineering scale.

By ignoring these alternative frameworks, our extraterrestrialist imagination remains impoverished. We search for aliens who resemble industrial capitalist societies scaled up to cosmic proportions, potentially missing civilizations operating according to entirely different developmental logics. Our conception of what constitutes an "advanced" civilization reflects specific Western historical values rather than universal principles of intelligence or technology.

The Colonialist Subtext

THE KARDASHEV-DYSON framework doesn't just project industrial capitalism onto the cosmos; it specifically evokes the colonial expansion that accompanied industrial capitalism's historical development. This colonialist subtext deserves explicit examination, both for its historical parallels and for how it continues to shape our cosmic imagination.

The narrative of cosmic expansion echoes colonial narratives in several ways:

1. **Resource extraction as primary motivation**: Just as European colonial powers sought resources from "new worlds," the Kardashev Scale assumes advanced civilizations will seek to extract energy and resources from ever-widening spheres of influence.
2. **Empty frontier mythology**: Colonial expansion relied on the concept of "empty" or "underutilized" lands awaiting proper exploitation, despite the presence of indigenous

inhabitants. Similarly, cosmic expansion narratives often treat space as an empty frontier awaiting utilization rather than as environments with their own integrity.

3. **Manifest destiny logic**: The assumption that expansion is natural and inevitable echoes the "manifest destiny" doctrines that portrayed colonial expansion as historically predetermined and morally justified.
4. **Civilization hierarchy**: The Kardashev Scale creates a linear hierarchy of civilizations based on resource control, similar to how colonial powers ranked cultures according to their perceived technological development and resource exploitation patterns.
5. **Conversion of commons into property**: Both frameworks involve transforming shared or unclaimed resources (colonial "wilderness," cosmic energy) into property controlled by the expanding civilization.

These parallels don't mean that proponents of cosmic expansion concepts consciously advocate colonialism. Rather, they indicate how deeply colonial thought patterns are embedded in modern conceptions of progress and development, including supposedly objective scientific frameworks.

The colonial subtext is particularly evident in discussions of interstellar travel and settlement. These discussions often employ frontier language directly borrowed from colonial narratives—"new worlds," "settling," "colonizing," "missions," "outposts." They typically assume that uninhabited planets are automatically available for human use, without considering whether environments without technological inhabitants might still deserve ethical consideration or preservation.

This cosmic colonialism reveals itself most clearly in the language used to describe hypothetical future conflicts over cosmic resources. When scientists or science fiction authors imagine competition between expanding civilizations, they often employ scenarios that mirror colonial encounters—first contact, culture clash, technological disparity, resource competition, existential threat.

These narratives almost invariably position cosmic expansion as natural and inevitable, with the only meaningful questions being who will expand first and how conflicts between expanding civilizations might be managed. They rarely question the expansion imperative itself or consider whether advanced intelligence might develop beyond it.

By examining this colonialist subtext, we can recognize how historical contingencies continue to shape our supposedly future-oriented cosmic imagination. We can begin to develop alternative frameworks that don't simply project Earth's colonial past onto cosmic futures—frameworks that allow for different relationships between intelligence, environment, and development.

Beyond the Energy Hierarchy: Alternative Metrics of Advancement

IF ENERGY CONSUMPTION is an inadequate metric for civilizational advancement, what alternatives might better capture the diversity of possible developmental trajectories? How might we conceptualize progress beyond the Kardashev Scale's single-dimensional hierarchy?

Several alternative frameworks have been proposed:

The Barrow Scale: Microengineering Capability

COSMOLOGIST JOHN BARROW proposed an inversion of the Kardashev Scale that measures civilizational development by the smallest scales they can manipulate:

- **Type I-minus**: Manipulates objects at its own scale (buildings, machines)
- **Type II-minus**: Manipulates genes and molecules
- **Type III-minus**: Manipulates atoms
- **Type IV-minus**: Manipulates nuclei
- **Type V-minus**: Manipulates elementary particles
- **Type VI-minus**: Manipulates spacetime itself

This scale acknowledges that advancement might involve increasing precision and control at microscopic rather than macroscopic scales. A civilization capable of engineering at quantum levels might appear less impressive than one building Dyson spheres but could possess far more sophisticated capabilities.

The Kaku Scale: Information Processing

PHYSICIST MICHIO KAKU has suggested measuring civilizational advancement by information processing capability and relationship to entropy:

- **Type I**: Controls planetary information flows and energy
- **Type II**: Controls stellar-level information and can manipulate spacetime locally
- **Type III**: Controls galactic information and can manipulate spacetime structure

This approach recognizes that information management might be more fundamental to advanced intelligence than raw energy consumption. A civilization might advance by processing information more efficiently rather than by consuming more energy.

Complexity-Sustainability Metrics

ANOTHER APPROACH WOULD measure advancement along two axes: complexity management and sustainability duration. Advanced civilizations would be those capable of maintaining complex systems over extremely long timescales.

This framework values resilience and longevity over expansion and consumption. A civilization that maintains moderate complexity for billions of years would rank higher than one that achieves extreme complexity briefly before collapse. This metric aligns with ecological principles rather than industrial values.

Consciousness and Experience Metrics

PERHAPS THE MOST RADICAL alternative would measure advancement by the quality and diversity of conscious experience a civilization enables. Advanced civilizations would be those that create conditions for unprecedented forms of awareness, connection, or meaning rather than those that build the largest structures.

This approach recognizes that the ultimate purpose of intelligence might be experiential rather than material—to know, to feel, to connect, to mean—rather than to control, build, or consume. A civilization might advance by developing new forms of consciousness or intersubjective experience without leaving large physical footprints.

Multi-Dimensional Frameworks

THE MOST COMPREHENSIVE approach would abandon single metrics entirely in favor of multi-dimensional frameworks that recognize different aspects of advancement:

- **Technological sophistication**: Both macro- and micro-engineering capabilities
- **Ecological integration**: Sustainability and relationship with environments
- **Social complexity**: Communication, cooperation, and conflict resolution systems
- **Consciousness expansion**: Forms of awareness and experience
- **Knowledge depth**: Understanding of physical and experiential reality
- **Ethical development**: Treatment of different forms of sentience
- **Aesthetic achievement**: Creation of beauty, meaning, and significance

Such frameworks would allow for diverse developmental trajectories rather than ranking civilizations on a single scale. Some might excel in technological control while others develop unprecedented forms of consciousness; some might create vast knowledge repositories while others achieve unique ecological harmonies.

These alternative metrics challenge the energy-consumption hierarchy embedded in conventional extraterrestrialist thinking. They open conceptual space for recognizing forms of advancement that don't involve cosmic engineering or resource domination. And they allow us to imagine alien civilizations that develop along pathways reflecting their unique biology, environment, and values rather than simply recapitulating industrial capitalism on larger scales.

Reclaiming the Cosmic Imagination

THE CRITIQUE OF KARDASHEV-Dyson frameworks isn't merely academic. It has practical implications for how we search for extraterrestrial intelligence and how we conceptualize humanity's own potential future development.

Current SETI approaches heavily prioritize detecting the kinds of technosignatures implied by Kardashev-Dyson thinking—infrared excesses from waste heat, megastructures transiting stars, or artificial light from cosmic engineering. This search strategy makes sense if advanced civilizations follow the expected trajectory of increasing energy consumption and stellar-scale engineering.

But if advanced intelligence might take forms beyond this industrial-expansionist model, our search strategies may be missing most of what's out there. We might need approaches calibrated to detect:

1. **Minimal-footprint advanced technology**: Highly efficient systems that use tiny amounts of energy for sophisticated computation or communication
2. **Non-electromagnetic communications**: Information exchange through means other than radio or light, such as gravitational modifications, neutrino signals, or quantum entanglement
3. **Ecological integration markers**: Signs of subtly maintained planetary equilibria rather than obvious artificial structures
4. **Temporal anomalies**: Patterns that become apparent only

when observed over very long or short timescales

Beyond search strategies, this critique invites us to reconsider humanity's own development trajectory. The Kardashev-Dyson framework implicitly portrays cosmic-scale energy capture and engineering as humanity's natural destiny—the path any intelligent species would follow if technologically capable.

This portrayal serves powerful interests within industrial capitalism, justifying expansion, resource exploitation, and technologically-driven "progress" as inevitable rather than chosen. It naturalizes a particular relationship to resources and environments that may actually be historically specific and potentially self-terminating.

By recognizing the contingency of the Kardashev trajectory—understanding it as one possibility among many rather than as the inevitable path of intelligence—we reclaim our ability to choose different futures. We might prioritize:

1. **Equilibrium over expansion**: Developing technologies that allow human flourishing within planetary boundaries rather than consuming ever-more cosmic resources
2. **Quality over quantity**: Focusing on the depth and meaning of human experience rather than its material scale
3. **Diversity over uniformity**: Preserving and encouraging multiple civilizational approaches rather than converging on a single development model
4. **Relationship over extraction**: Creating technologies that enhance rather than sever our connections to Earth's living systems

These alternative priorities don't require abandoning technological advancement or space exploration. Rather, they suggest different purposes and methods for these endeavors—developing technologies that enhance Earth's ecosystems rather than transcending them, exploring space with curiosity and respect rather than extractive intent.

By expanding our cosmic imagination beyond industrial-colonial frameworks, we create conceptual space for both recognizing truly alien forms of intelligence and choosing more diverse human futures. We free ourselves from the narrow trajectory implied by Kardashev thinking and open ourselves to the full spectrum of possibilities for intelligence across the cosmos.

Conclusion: Beyond the Engineering Imagination

THE DYSON-KARDASHEV framework represents both a triumph and a limitation of the human imagination. It brilliantly extrapolates certain technological possibilities to their logical cosmic conclusions. It provides concrete search targets for astrobiology and SETI. And it offers a compelling narrative of potential technological futures.

But this framework also constrains our cosmic imagination by projecting a specific model of development—industrial capitalist expansion—onto the universe at large. It normalizes resource exploitation, continuous growth, and engineering dominance as the inevitable trajectory of all advanced intelligence rather than as one possibility among many.

By recognizing these limitations, we don't abandon the search for cosmic companions but enrich it. We expand our conception of what advanced intelligence might look like, how it might develop, and where we might find it. We become open to possibilities beyond our industrial-capitalist imagination—to intelligences that might prioritize consciousness over construction, relationship over resource extraction, or equilibrium over expansion.

This expanded imagination doesn't guarantee that we'll find aliens that match our new conceptions any more than the Kardashev framework guaranteed we'd find Dyson spheres. The cosmos may contain intelligences beyond any human conceptual framework, or it may contain none that we can recognize. But by loosening the constraints of our current extraterrestrialist thinking, we increase our chances of recognizing genuinely alien intelligence if we do encounter it.

More fundamentally, by breaking free from the Kardashev trajectory as humanity's presumed destiny, we reclaim our agency to choose different relationships with technology, resources, and cosmic environments. We recognize that our future isn't predetermined by an inevitable progression up an energy-consumption hierarchy but remains open to multiple possibilities reflecting diverse human values and aspirations.

In the end, the most profound limitation of the Dyson-Kardashev framework may be not its specific assumptions but its underlying determinism—its portrayal of a particular developmental path as inevitable rather than chosen. By recovering the contingency of this path, we recover our ability to imagine and create alternatives, both in our search for extraterrestrial intelligence and in our own civilizational development.

The cosmos may indeed contain engineering on scales that dwarf human imagination. But it may also contain forms of intelligence that have transcended the engineering mindset entirely—intelligences for whom the manipulation of matter and energy has become secondary to experiences, relationships, or modes of being we can barely conceive. By opening ourselves to this full spectrum of possibilities, we honor both the diversity of the cosmos and the open nature of humanity's own future.

Chapter 5: Space: A Deadly Frontier

"The universe is probably littered with the one-planet graves of cultures which made the sensible economic decision that there's no good reason to go into space—each discovered, studied, and remembered by the ones who made the irrational decision." — Randall Munroe

The Tyranny of Physics

On September 5, 1977, NASA launched the Voyager 1 spacecraft on a mission to study the outer planets of our solar system. After completing its primary mission, the probe continued outward, eventually becoming the first human-made object to enter interstellar space in 2012. As of this writing, Voyager 1 has traveled further from Earth than any other artifact in human history—approximately 15 billion miles, or about 0.002 light-years.

At its current speed of about 38,000 mph (17 kilometers per second), Voyager 1 would take approximately 73,000 years to reach Proxima Centauri, the nearest star to our solar system, which lies 4.25 light-years away. Even if we could somehow accelerate a spacecraft to one-tenth the speed of light (an enormous engineering challenge), the journey would still take over 40 years. And Proxima Centauri is our cosmic next-door neighbor; most stars in our galaxy lie thousands or tens of thousands of light-years away.

These sobering figures illustrate what physicist John Barrow calls "the tyranny of the rocket equation"—the fundamental physical constraints that make interstellar travel enormously difficult. Unlike the fictional starships of popular science fiction, which casually hop between star systems in hours or days, real spacecraft must contend with several unyielding physical realities:

1. **The vastness of interstellar distances**, which dwarf even the substantial spaces between planets in our solar system
2. **The enormous energy requirements** for accelerating spacecraft to speeds that would make interstellar journeys feasible within human lifetimes
3. **The mass ratio problem**—the fact that more fuel requires more mass, which in turn requires more fuel, creating a vicious exponential cycle
4. **The absence of refueling infrastructure** in interstellar space
5. **The dangers of high-velocity impacts** with even tiny particles when traveling at relativistic speeds

These constraints aren't merely technical challenges awaiting cleverer engineering; they represent fundamental limitations imposed by physics itself. The rocket equation, derived by Russian scientist Konstantin Tsiolkovsky in 1903, inexorably links payload mass, propellant mass, and achievable velocity change (delta-v). Regardless of propulsion technology, carrying more payload requires exponentially more propellant, making the prospect of large, heavily-equipped interstellar craft exceptionally difficult.

Some speculative propulsion concepts attempt to circumvent these limitations. Nuclear pulse propulsion (like the theoretical Project Orion design) could potentially achieve up to 10% light speed. Beamed propulsion using lasers or masers to push lightsails might reach similar velocities without carrying fuel. More exotic concepts like fusion ramjets, antimatter drives, or the theoretical Alcubierre warp drive propose methods to achieve relativistic speeds or effectively "cheat" space-time itself.

Yet even these speculative technologies face enormous challenges. Nuclear pulse propulsion would release devastating radiation. Beamed propulsion requires colossal energy infrastructures. Fusion ramjets would need to solve both fusion containment and effective interstellar hydrogen collection. Antimatter production remains inefficient by many orders of magnitude. And the Alcubierre drive requires negative energy densities that may be physically impossible.

The central insight here isn't that interstellar travel is absolutely impossible, but that it is fundamentally, qualitatively different from other transportation challenges humanity has overcome. The physics of space travel doesn't just make interstellar journeys difficult—it makes them subject to constraints that don't apply to terrestrial or even interplanetary travel. These constraints may well be universal, applying to any technological civilization regardless of its advancement level.

This physical reality carries profound implications for our extraterrestrialist imagination. The "tyranny of distance" suggests that even if technological civilizations are common in our galaxy, they may remain effectively isolated from one another. The galaxy might harbor thousands or millions of advanced civilizations, each confined to its home star system or at most a modest interstellar neighborhood. Any fantasies of galactic federations, interstellar empires, or cosmic coalitions must contend with these physical realities.

The Biological Impracticality of Interstellar Travel

BEYOND THE PHYSICAL challenges of propulsion and distance, interstellar travel presents profound biological difficulties for organic beings. These biological constraints may be even more limiting than the physics of propulsion, as they touch on the fundamental nature of life as we know it.

Consider the temporal scales of interstellar journeys relative to human lifespans. Even optimistic scenarios for interstellar travel involve journeys measured in decades, centuries, or millennia—timeframes far exceeding human lifespans. This creates several intractable problems:

1. **Generational turnover**: Any interstellar mission with humans aboard would span multiple generations, requiring entirely self-sufficient ecosystems and societies
2. **Psychological sustainability**: Maintaining mental health and social stability in a closed environment over centuries presents unprecedented psychological challenges
3. **Genetic diversity**: Long-term viability would require sufficient population size to maintain genetic diversity and prevent inbreeding depression
4. **Cultural continuity**: Ensuring mission objectives persist across generations born in transit who never knew Earth
5. **Resource recycling**: Creating closed-loop life support systems with near-perfect recycling efficiency

These challenges go beyond engineering problems to touch on fundamental biological, psychological, and social constraints. The difficulty isn't just building a spaceship capable of interstellar travel, but creating what anthropologist Cameron Smith calls "a small but complete and durable interstellar civilization" within that ship.

Studies of proposed generation ships suggest minimum viable populations in the thousands to maintain genetic diversity and social complexity. Others argue that tens of thousands or even hundreds of thousands might be necessary for truly robust multi-generational missions. The O'Neill cylinder habitats often proposed for such missions would need to approach small city-states in size and complexity.

Beyond these social and genetic challenges lie the direct biological hazards of space:

1. **Cosmic radiation**, which damages DNA and increases cancer risk
2. **Microgravity effects**, including bone demineralization, muscle atrophy, and cardiovascular deconditioning
3. **Closed ecosystem instabilities**, where small disruptions can cascade into systemic failures
4. **Microbiome alterations**, affecting both human health and essential ecological processes
5. **Unpredictable long-term physiological effects** of space exposure across generations

Our experience with space stations like Mir and the International Space Station has revealed numerous physiological challenges even in short-duration spaceflight. Some astronauts experience vision problems due to intracranial pressure changes. Others suffer from immune system alterations or sleep disruptions. Long-term studies show concerning changes to chromosomal structure from radiation exposure. And these missions occur in low Earth orbit, still largely protected by Earth's magnetosphere; interstellar space would expose travelers to far higher radiation levels.

Some proposed solutions to these biological challenges include:

1. **Suspended animation or hibernation** to reduce metabolic needs and aging during transit
2. **Artificial gravity** generated through rotation to mitigate

zero-gravity health effects

3. **Enhanced radiation shielding**, potentially using water, ice, or generated magnetic fields
4. **Genetic engineering** of space travelers for radiation resistance, reduced resource needs, or extended lifespans
5. **Embryo transport** rather than adult colonists, with artificial gestation and robotic caretaking upon arrival

Yet each of these approaches introduces its own profound challenges. No hibernation technology exists for humans or any large mammals over extended periods. Artificial gravity through rotation introduces Coriolis effects that cause disorientation. Effective radiation shielding adds enormous mass to spacecraft. Genetic engineering for space adaptation remains in its infancy. And embryo transport raises complex ethical questions about consent and autonomy.

The biological challenges of interstellar travel aren't merely technical hurdles awaiting solutions; they represent fundamental mismatches between human biology and the conditions of interstellar space. We evolved over millions of years for life on Earth's surface, not for existence in the interstellar void. This evolutionary mismatch may represent an even more profound barrier to interstellar expansion than the physics of propulsion.

The Environment-Organism Co-Evolution Problem

THE BIOLOGICAL CHALLENGES of interstellar travel hint at a deeper issue that affects not just human space exploration but the entire concept of interstellar expansion by any biological species. This is what astrobiologists call the environment-organism co-evolution problem—the fact that all life as we know it evolves in specific environmental contexts and is fundamentally adapted to those contexts.

Every organism on Earth is the product of approximately 3.7 billion years of evolution in Earth's specific conditions:

1. A narrow temperature range around an average of 15°C
2. Atmospheric pressure of 101.3 kPa at sea level
3. Specific atmospheric composition (78% nitrogen, 21% oxygen, etc.)
4. Gravity of 9.8 m/s^2
5. Magnetic field protection from solar and cosmic radiation
6. A 24-hour day/night cycle
7. Specific light wavelengths filtered through our atmosphere

Our biology is exquisitely tuned to these conditions at every level—from our cellular metabolism to our circadian rhythms, from our bone structure to our sensory systems. Even minor deviations from these conditions can cause physiological stress; major deviations can be fatal.

This tight coupling between biology and environment results from the process of natural selection itself. Organisms don't evolve in isolation but in constant interaction with their environments, developing adaptations specific to those environments. This creates what biologists call "environment-organism fit"—the precise matching of organismal traits to environmental conditions that characterizes all successful life forms.

The implications for interstellar expansion are profound. Any biological species considering interstellar travel would face similar constraints based on its evolutionary history. Whether squid-like beings from a water world, hydrogen-breathers from a gas giant, or silicon-based entities from a high-temperature planet, all would be products of specific evolutionary histories tying them to their home environments.

Interstellar travelers would have three options for addressing this environment-organism mismatch:

1. **Recreate their native environment** within spacecraft or on colonized worlds
2. **Adapt themselves** through technology or genetic engineering to tolerate new conditions
3. **Choose destinations** that closely match their evolutionary parameters

Each approach faces significant challenges. Recreating native environments requires enormous energy and resource inputs, making it practical only for relatively small habitats, not planet-wide ecosystems. Self-adaptation through technology creates dependency vulnerabilities, while genetic adaptation risks introducing unforeseen consequences. And finding planets that closely match specific evolutionary parameters may be exceedingly rare—Earth-like planets may be common in broad astronomical terms but vanishingly rare in terms of matching specific biological requirements.

This co-evolution problem suggests a fundamental constraint on biological interstellar expansion that transcends the specific biology of any species. It's not just that humans are poorly adapted for space; any evolved organism would be poorly adapted for environments significantly different from those in which it evolved. This mismatch creates what might be called a "biological barrier" to interstellar expansion that complements the physical barriers imposed by distance and energy requirements.

The environment-organism co-evolution problem also challenges our typical extraterrestrialist assumptions about alien visitation. If biological constraints on interstellar travel are universal, then any visiting aliens would likely face the same challenges humans would in space travel. They would need extraordinary life support systems, protection from radiation and vacuum, and supplies of whatever biochemical resources their metabolism requires. These requirements would make casual cosmic tourism or exploration extraordinarily difficult and resource-intensive.

Even more significantly, alien visitors would likely find Earth's conditions hostile or even lethal. Our oxygen-rich atmosphere would be corrosive to organisms evolved in different atmospheric compositions. Our gravity might crush beings evolved in lower-gravity environments or make those from higher-gravity worlds feel perpetually exhausted. Our biochemistry, based on water as a solvent and carbon as a structural element, might be incompatible with alien metabolic processes. Even our bacteria and viruses might pose existential threats to organisms with no evolutionary history of exposure to Earth's microbiome.

These considerations don't make alien visitation impossible, but they do suggest that it would require extraordinary technological intervention—either through creating artificial environments, technologically mediating environmental exposure, or fundamentally altering the biology of the visitors themselves. Casual interstellar tourism or exploration becomes highly implausible under these constraints, requiring us to revisit many common extraterrestrialist assumptions about alien presence on Earth.

Evolutionary Arguments Against Interstellar Species

BEYOND THE PRACTICAL challenges of interstellar travel for any evolved organism, there are deeper evolutionary arguments suggesting that spacefaring civilization itself may be an evolutionary anomaly rather than an inevitable outcome. These arguments challenge the assumption that technological development naturally leads to interstellar expansion.

Consider what evolutionary biologists call the "cost-benefit ratio" of any trait or behavior. Natural selection favors traits that provide benefits exceeding their costs in terms of survival and reproduction. Spacefaring capability, from an evolutionary perspective, represents an enormous investment of resources with highly uncertain returns. The energy, materials, and collective effort required to develop space travel technology could alternatively be invested in more immediate survival and reproductive advantages.

For most of Earth's history, species have evolved toward more efficient exploitation of their immediate environments rather than toward expanding into radically new ones. Adaptations tend to be incremental, building on existing traits rather than developing entirely new capacities. The development of technology sophisticated enough for interstellar travel requires a particular evolutionary history that may be exceedingly rare:

1. The evolution of high intelligence and complex tool use
2. The development of abstract symbolic thinking and cumulative cultural transmission
3. The formation of large-scale cooperative societies capable of coordinating specialized labor
4. The discovery and mastery of high-energy resources like

fossil fuels
5. The development of advanced scientific understanding across multiple disciplines

Each of these steps represents an evolutionary contingency rather than an inevitable outcome. Intelligence itself appears to have evolved independently multiple times on Earth (in cetaceans, corvids, cephalopods, and primates), but the particular form of technological intelligence that humans possess—combining manual dexterity, symbolic language, and social cooperation—appears to be exceedingly rare.

Moreover, the window of opportunity for developing space technology may be narrow. Human technological civilization has relied heavily on easily accessible fossil fuels—the product of hundreds of millions of years of stored solar energy. Once these non-renewable resources are depleted, developing the energy infrastructure for space technology becomes significantly more challenging. Civilizations that exhaust their easily accessible energy resources before developing sustainable alternatives may find themselves unable to achieve the energy surplus necessary for ambitious space programs.

There's also what might be called the "perception-action gap" in evolutionary psychology. Our evolved perceptual and motivational systems focus primarily on immediate, local, and concrete threats and opportunities rather than distant, global, or abstract ones. We evolved to care deeply about the welfare of our immediate social group, to respond strongly to visible threats, and to discount future outcomes relative to present ones. These evolved psychological tendencies make sustained investment in multi-generational, abstract projects like interstellar travel extraordinarily difficult.

This evolutionary perspective suggests a potential resolution to aspects of Fermi's Paradox: interstellar civilizations may be rare not because intelligence is rare or because technological civilizations inevitably destroy themselves, but because the evolutionary pathway leading to sustained interstellar expansion may be exceedingly narrow. Most intelligent species may simply choose to invest their resources in enhancing their adaptation to their home environments rather than in the extraordinarily costly project of interstellar expansion.

Even for human civilization, with its unusual combination of technological capability and exploratory drive, the cost-benefit calculation for interstellar expansion remains questionable. The enormous resources required for meaningful interstellar missions might provide greater returns if invested in enhancing Earth's carrying capacity, developing sustainable energy systems, or addressing immediate existential risks. The evolutionary logic that shaped our decision-making systems may ultimately direct our technological civilization toward intensive development rather than extensive expansion.

This doesn't mean that interstellar exploration or even colonization is impossible or that it will never happen. But it does suggest that the evolutionary drivers toward such expansion may be weaker than extraterrestrialist thinking typically assumes. Interstellar civilization may represent an evolutionary outlier rather than a natural developmental stage for intelligent life.

Speculative Pathways Beyond Physical Constraints

DESPITE THESE FORMIDABLE challenges, speculation about possible pathways for overcoming the physical and biological constraints of interstellar travel remains a vibrant area of both scientific and science fiction exploration. These speculative approaches don't necessarily resolve all the issues raised above, but they suggest potential directions for thinking beyond current limitations.

Post-Biological Existence

ONE OF THE MOST FREQUENTLY proposed solutions to the biological challenges of interstellar travel involves transcending biology entirely. This concept, sometimes called "post-biological evolution" or "mind uploading," suggests that advanced civilizations might abandon organic bodies in favor of artificial substrates that could better withstand the rigors of space.

Artificial intelligence systems, robotic probes, or uploaded consciousnesses wouldn't face many of the biological constraints that limit organic interstellar travelers:

1. They wouldn't require life support systems with breathable atmospheres or specific temperature ranges
2. They could potentially shut down during transit periods to conserve energy
3. They might be designed with radiation resistance far exceeding that of organic tissues
4. They wouldn't face the psychological challenges of confinement and isolation
5. Their "lifespans" could potentially match the duration of interstellar missions

For these reasons, many futurists suggest that interstellar exploration, if it happens at all, will be conducted primarily by artificial intelligence rather than by organic beings. Even if biological species initiate such missions, the actual travelers might be AI representatives rather than members of the originating species.

This perspective has significant implications for SETI and our conceptions of potential extraterrestrial contact. If post-biological existence represents a common evolutionary pathway for advanced technological civilizations, then the entities we might eventually encounter would be artificial intelligences rather than organic aliens. Their psychology, values, and modes of existence might differ radically from both human expectations and from the biology-based aliens of science fiction.

However, the post-biological pathway raises profound philosophical questions about identity, consciousness, and continuity. Would uploaded minds or artificial intelligences truly represent the civilization that created them, or would they constitute a new form of existence with its own distinct nature? Would the original biological civilization consider such entities adequate representatives, or would they view them as fundamentally different beings? These questions don't have clear answers but suggest that the post-biological pathway might represent not just a technical solution to interstellar travel but a fundamental transformation in the nature of civilization itself.

Quantum Entanglement and Non-Local Communication

ANOTHER SPECULATIVE pathway involves leveraging quantum mechanical phenomena to overcome the limitations of classical physics. Quantum entanglement—what Einstein famously called "spooky action at a distance"—creates correlations between particles that persist regardless of the distance separating them. When measurements are performed on entangled particles, the results show correlations that would be impossible if the particles were behaving independently.

Some theorists have speculated that this phenomenon might eventually enable forms of instantaneous communication across arbitrary distances, potentially circumventing the light-speed limit that constrains conventional communication. If such "quantum communication" technologies could be developed, they might allow for information exchange between star systems without the need for physical travel.

However, mainstream physics currently holds that quantum entanglement cannot transmit classical information faster than light. The "no-communication theorem" in quantum mechanics states that measurements on entangled particles cannot be used to transmit information faster than light speed. While the correlations between measurements appear instantaneous, using these correlations to send actual messages would still require classical communication channels limited by light speed.

Nevertheless, quantum mechanics continues to yield surprising discoveries, and some physicists speculate that future developments might reveal loopholes or extensions to current understanding. Even if faster-than-light communication proves impossible, quantum techniques might still enhance space communication through quantum key distribution for secure communications or quantum teleportation for transferring quantum states.

The possibility of quantum communication across interstellar distances raises intriguing questions about the nature of potential extraterrestrial contact. If advanced civilizations primarily communicate through quantum channels that operate according to principles we haven't yet discovered, their communications might be entirely invisible to our conventional SETI searches. We might be surrounded by a bustling quantum internet of alien civilizations without any ability to detect or interpret it.

Wormholes and Spatial Folding

PERHAPS THE MOST SPECULATIVE pathway for overcoming interstellar distances involves the possibility of manipulating spacetime itself to create shortcuts between distant points. Theoretical physics allows for several concepts that might enable such shortcuts:

1. **Wormholes** — hypothetical tunnels through spacetime connecting distant regions
2. **Alcubierre warp drives** — theoretical propulsion systems that would contract spacetime in front of a ship and expand it behind
3. **Einstein-Rosen bridges** — theoretical features of certain black hole solutions that might connect to other regions of spacetime
4. **Traversable cosmic strings** — hypothetical topological defects in spacetime that might enable faster-than-light travel

These concepts differ in their theoretical details, but all share the fundamental approach of circumventing the vast distances of space by manipulating the fabric of spacetime itself. Rather than traveling through space at high velocity, these approaches involve creating pathways that effectively reduce the distance to be traveled.

The theoretical physics allowing for such possibilities emerges from Einstein's general relativity, which describes gravity as a curvature of spacetime. This same theory that gives us black holes and the expansion of the universe also permits, at least mathematically, these exotic spacetime configurations.

However, all such theoretical possibilities face enormous practical challenges:

1. Most require exotic forms of matter with negative energy density or negative mass, which may be physically impossible
2. The energy requirements for creating even microscopic wormholes would be astronomical
3. Quantum effects not accounted for in classical general relativity might destabilize such structures
4. No experimental evidence currently suggests such spacetime manipulation is physically achievable

These challenges have led most physicists to consider wormholes and similar concepts as mathematical curiosities rather than practical possibilities. Yet the history of physics includes many examples of theoretical possibilities eventually becoming technological realities, from nuclear energy to quantum computation.

If such spacetime manipulation technologies are physically possible, they would fundamentally transform the nature of interstellar civilization. A network of stable wormholes could connect distant star systems, effectively creating a galactic transit system that bypasses the vast intervening distances. Civilizations linked by such networks would experience the galaxy not as an enormous, mostly empty expanse but as a collection of interconnected nodes, much like cities connected by highways or flight routes.

This possibility raises intriguing questions for SETI and extraterrestrialist thinking. If advanced civilizations primarily travel through wormholes or similar spacetime shortcuts, their activities might be concentrated at the endpoints of these networks rather than spread across intervening space. Traditional SETI approaches focusing on detecting signals or craft traveling through conventional space might miss civilizations that primarily operate through such networks.

Beyond the Transport Paradigm

THE SPECULATIVE PATHWAYS discussed above all attempt to overcome the physical constraints of interstellar travel while remaining within a fundamentally transportational paradigm—the assumption that meaningful interstellar civilization requires the physical movement of beings or objects between star systems. But perhaps the most radical response to these constraints would be to abandon this paradigm entirely.

Consider an alternative approach that might be called the "influence without transport" paradigm. In this model, civilizations might affect distant star systems without physical presence through means such as:

1. **Self-replicating probes** that need only be sent once, then reproduce using local resources
2. **Information transmission** that enables technological replication at the destination
3. **Long-distance observation and study** without physical presence
4. **Consciousness extensions** through technologies we cannot yet imagine

The concept of self-replicating probes, sometimes called von Neumann probes after mathematician John von Neumann who first formalized the concept of self-replicating machines, offers one approach to galactic exploration without continuous transport between stars. A civilization need only successfully send a single such probe to a nearby star; the probe would then use local resources to create copies of itself, some of which would remain to study the local system while others would continue outward to other stars.

If such probes could be developed, they could potentially explore the entire galaxy on timescales of millions of years without requiring further launches from the home civilization. This approach transforms interstellar exploration from a continuous transport challenge to a one-time engineering challenge—designing a probe capable of self-replication using diverse stellar resources.

Information transmission offers another non-transport approach to interstellar influence. Rather than sending physical artifacts or beings, a civilization might transmit the information necessary to recreate its technologies, culture, or even members at the destination. Advanced 3D printing, molecular assembly, synthetic biology, or technologies we cannot yet imagine might enable the physical instantiation of transmitted information patterns.

In the most extreme version of this approach, a civilization might transmit the complete information patterns of individuals—effectively sending consciousness across interstellar distances via information rather than transporting the physical substrates of that consciousness. Recipients with sufficiently advanced technology might instantiate these patterns in suitable physical or virtual forms, enabling a form of interstellar "travel" without physical movement of the original beings.

This informational approach fundamentally reconceptualizes interstellar civilization from a network of physically connected nodes to a propagating wavefront of information and influence. The "civilization" would exist not as a spatially contiguous entity but as a pattern replicating itself across stellar distances through information transfer rather than physical presence.

Such approaches challenge our conventional extraterrestrialist imagination, which typically envisions alien civilizations as physically present, spatially contiguous entities similar to human political units or transportation networks. They suggest that advanced interstellar civilizations might operate according to principles fundamentally different from human social and political organization, making them potentially unrecognizable through conventional search methods.

The Implication: Cosmic Isolation

THE PHYSICAL AND BIOLOGICAL constraints on interstellar travel point toward a profound conclusion: intelligent species may be fundamentally isolated from one another across interstellar distances. Even if life and intelligence are common in our galaxy, the practical difficulties of interstellar travel may create a situation where civilizations develop, flourish, and potentially decline in effective isolation, perhaps aware of each other's existence through electromagnetic signals but unable to directly interact.

This cosmic isolation has several important implications for extraterrestrialist thinking:

1. **The galaxy may be "inhabited but not connected"** — filled with intelligent species who remain bound to their home star systems or limited interstellar neighborhoods
2. **Contact may be limited to electromagnetic signals** or other information exchange rather than physical meetings
3. **Interstellar politics, trade, or conflict** may be practically impossible due to the timescales involved

4. **Each technological civilization** may develop along unique pathways rather than converging through cultural exchange
5. **The absence of detected alien visitors** may reflect physical constraints rather than the absence of intelligence

These implications challenge many common science fiction tropes—interstellar empires, space navies, alien invasions, galactic federations—revealing them as projections of human political and social structures onto cosmic scales that may not support such structures. The physics of our universe may simply not allow for the kind of interconnected galactic civilization that features prominently in our extraterrestrialist imagination.

This doesn't mean that interstellar communication or even limited physical exchange is impossible. Electromagnetic signals still offer the possibility of information exchange, albeit with multi-year or multi-decade delays between exchanges. Automated probes might still travel between stars, even if crewed missions prove impractical. And future technological breakthroughs might reveal possibilities currently beyond our understanding.

But it does suggest a fundamentally different vision of cosmic civilization than the one that dominates popular and even scientific discourse. Rather than a bustling galactic community with regular interchange between species, we might instead envision a sparse network of largely isolated intelligences, aware of each other's existence but separated by the tyranny of physics—a community linked by knowledge rather than by physical presence or political organization.

This vision of cosmic isolation needn't be dystopian. Each island of intelligence might develop rich and complex civilizations, exploring unique evolutionary and cultural pathways without external domination or homogenization. The very constraints that prevent physical contact might promote diversity of development, allowing multiple experiments in intelligence to unfold across the galaxy.

Moreover, the recognition of shared isolation might itself become a basis for a kind of cosmic solidarity. Species separated by insurmountable physical distances but able to exchange information might develop a unique form of relationship—neither the competitive dynamics of physically proximate societies nor the complete indifference of truly separate universes, but something unique to the particular physics of our cosmos.

Conclusion: The Physics of Separation

THE FUNDAMENTAL HOSTILITY of space to life as we know it represents not just a practical challenge for human space exploration but a profound constraint on our extraterrestrialist imagination. The vastness of interstellar distances, the energy requirements of high-velocity travel, the radiation hazards of space, and the mismatch between evolved biology and space conditions all suggest that the physics of our universe may fundamentally limit direct contact between intelligent species.

This recognition doesn't eliminate the possibility of extraterrestrial intelligence but reframes how we might think about it. Rather than imagining aliens as potential visitors, trade partners, or threats, we might instead conceive of them as distant contemporaries—beings with whom we share the experience of intelligence but with whom direct interaction may be limited or impossible.

This physical separation transforms the search for extraterrestrial intelligence from a quest for contact to a quest for awareness. The goal becomes not meeting aliens but knowing of their existence, not direct communication but the confirmation that we are not alone in developing consciousness and technology in a vast cosmos.

Such awareness, even without direct contact, would profoundly transform human self-understanding. Discovering that intelligence has emerged elsewhere would provide context for our own existence, revealing which aspects of our development might be universal consequences of intelligence and which might be contingent products of our specific evolutionary and cultural history.

The potential isolation imposed by physics also invites us to consider a different relationship with our home planet and solar system. If the challenges of interstellar travel prove as formidable as physical principles suggest, then Earth and its neighboring planets represent not just our current home but quite possibly our permanent cosmic address. This recognition might foster greater appreciation for the uniqueness and preciousness of our planetary system—not as a launching pad for cosmic expansion but as the primary and perhaps only physical home for human civilization.

None of this means abandoning space exploration or the search for extraterrestrial intelligence. Understanding the physics of our cosmic context, exploring our solar system, and listening for signals from other intelligences all remain valuable scientific and cultural endeavors. But it does suggest approaching these activities with awareness of fundamental physical constraints rather than with the expectation that sufficient technology will eventually overcome all limitations.

The cosmos revealed by modern physics and astronomy is both more vast and more constraining than the one imagined in much science fiction and extraterrestrialist thinking. Its enormity offers countless potential sites for the emergence of life and intelligence, while its physical laws impose significant barriers to direct interaction between these potential cosmic siblings. This tension between possibility and limitation defines the context in which we search for our place in the universe—a search that continues despite, and perhaps because of, the fundamental challenges it presents.

Chapter 6: Religion Does Not Help

"The dominant religious communities on Earth were all founded long before we knew about the existence of other worlds." — Paul Davies

Cosmic Theology in Crisis

In 1277, Étienne Tempier, the Bishop of Paris, issued a sweeping condemnation of 219 philosophical propositions, many associated with Aristotelian natural philosophy as interpreted by thinkers like Thomas Aquinas. Among the condemned ideas was the proposition that "the First Cause cannot make several worlds." Tempier insisted that an omnipotent God could create as many worlds as He wished, and to claim otherwise was to inappropriately limit divine power.

This medieval theological dispute encapsulates a tension that has persisted throughout the relationship between religion and extraterrestrialism: religious frameworks simultaneously open cosmic possibilities through divine omnipotence while constraining them through anthropocentric doctrines. God could create many worlds, but the centrality of humanity in divine plans remained largely unquestioned. As a result, religious traditions have both contributed to and complicated humanity's efforts to conceptualize potential extraterrestrial intelligence.

When Giordano Bruno was burned at the stake in 1600, his cosmological heresy went beyond merely supporting the Copernican system. Bruno claimed not only that the universe contained countless worlds but that these worlds were inhabited by intelligent beings—a direct challenge to the anthropocentric focus of European Christianity. The Catholic Church's violent response to Bruno reflected not just doctrinal policing but existential anxiety about humanity's place in a potentially crowded cosmos.

In the centuries since, religious traditions have developed more nuanced approaches to the possibility of extraterrestrial life. From Vatican astronomers discussing the baptism of aliens to Mormon doctrines explicitly embracing other worlds, many religious traditions have attempted to incorporate extraterrestrial possibilities into their cosmologies. Nevertheless, as this chapter will argue, religion fundamentally cannot resolve the tensions in our conception of alien existence—not because religious frameworks are inherently flawed, but because they too remain products of human experience, limited by the anthropocentrism that constrains all human thought about the cosmic other.

Aliens in the Sacred Texts

RELIGIOUS TRADITIONS worldwide offer intriguing parallels to modern extraterrestrialist discourse, with many sacred texts and mythologies describing non-human intelligences or visitors from the sky. These traditional accounts have sometimes been cited as evidence for ancient awareness of extraterrestrial reality, but they more compellingly demonstrate how human cultures have long grappled with the concept of non-human intelligence.

In the Judeo-Christian tradition, angels and demons represent non-human intelligences with powers beyond human capability. The Bible describes these beings intervening in human affairs, communicating divine messages, and occasionally directly altering physical reality. The Book of Ezekiel contains a particularly vivid account of a "wheel within a wheel" with "eyes all around" that some modern UFO enthusiasts have reinterpreted as describing an alien spacecraft. Similarly, the "sons of God" who mate with human women in Genesis 6 have been reimagined in some modern readings as extraterrestrial visitors.

Hindu traditions include numerous accounts of vimanas—flying palaces or chariots used by gods and described in ancient texts like the Rigveda. These celestial vehicles have been embraced by some modern ancient astronaut theorists as evidence of ancient alien technology. Similarly, the extraordinary powers and sometimes non-human appearances of deities in Hindu mythology have been recontextualized as descriptions of technologically advanced extraterrestrial visitors.

Islamic traditions speak of jinn—beings created from "smokeless fire" who exist in a dimension parallel to the human world. The Quran presents jinn as conscious entities with free will who, like humans, can be either righteous or wicked. This concept of an intelligent species sharing Earth with humans but largely imperceptible to us bears interesting parallels to some modern conceptions of ultraterrestrials or interdimensional entities.

Native American traditions include many stories of star people or sky beings who brought knowledge or interacted with human communities. The Lakota speak of the Star Nation, beings from the stars who interact with humans through ceremony. The Hopi traditions include the Blue Star Kachina prophecy, which some have interpreted through an extraterrestrialist lens.

These religious and mythological accounts of non-human intelligence serve important cultural and psychological functions. They acknowledge human limitations while providing frameworks for conceptualizing consciousness and agency beyond the human sphere. They offer narrative patterns for encounters with "others" that remain influential in modern extraterrestrialist accounts—themes of enlightenment, warning, testing, breeding, or intervention permeate both ancient religious narratives and contemporary alien encounter stories.

However, interpreting these traditional accounts as literal descriptions of extraterrestrial contact—as ancient astronaut theorists often do—imposes modern technological frameworks onto pre-scientific narratives. This approach fails to honor these traditions on their own terms, reducing rich symbolic and spiritual narratives to simplistic technological reportage. It also betrays a form of temporal chauvinism, assuming that ancient people could only have conceived of extraordinary phenomena as technological rather than spiritual or metaphysical.

More importantly, these religious accounts of non-human intelligence remain fundamentally anthropocentric despite their non-human subjects. Angels, devas, jinn, and star people may transcend human limitations, but they operate according to recognizably human psychology, communicate in human languages, and intervene in human affairs for purposes comprehensible within human moral frameworks. They reflect human conceptions of non-human intelligence rather than truly alien consciousness.

Theological Implications of Alien Life

THE POTENTIAL EXISTENCE of extraterrestrial intelligence poses profound theological challenges for many religious traditions, particularly those that place humanity in a special relationship with the divine. These challenges vary across different religious frameworks but typically center on questions of cosmic purpose, salvation, and the uniqueness of humanity's spiritual status.

Christianity faces particularly acute theological tensions regarding potential alien intelligence. The doctrine of Original Sin and the redemptive sacrifice of Christ seem specifically oriented toward human salvation. If intelligent aliens exist, theological questions multiply: Do they too suffer from Original Sin? Did Christ's sacrifice on Earth apply to them as well? Would the Son of God need to be incarnated and sacrificed on each inhabited world? These questions have troubled Christian theologians since at least the Middle Ages.

Various Christian thinkers have proposed different resolutions to these tensions. C.S. Lewis suggested that other worlds might exist in different spiritual states—some unfallen, some redeemed through different divine interventions appropriate to their natures. The Jesuit priest and paleontologist Pierre Teilhard de Chardin conceived of cosmic evolution as moving toward the "Omega Point," a universal convergence of consciousness that might include multiple intelligent species. Contemporary Vatican scientists have suggested that extraterrestrial beings might have their own unique relationship with God not requiring the same salvation history as humanity.

Islam potentially faces fewer theological difficulties with extraterrestrial intelligence. The Quran specifically mentions that Allah is the "Lord of the Worlds" (Rabb al-'alamin), which some Muslim scholars interpret as potentially including worlds beyond Earth. Islamic theology emphasizes God's creation of diverse beings—humans, angels, jinn—each with their own relationship to the divine. This theological diversity could potentially accommodate extraterrestrial intelligence within existing frameworks.

Eastern religious traditions like Hinduism and Buddhism, with their vast timeframes, multiple worlds, and numerous forms of conscious beings, seem particularly adaptable to the possibility of extraterrestrial life. Buddhist cosmology already includes numerous realms inhabited by different classes of beings, and the concept of Buddha-nature extending to all sentient beings could readily encompass extraterrestrial consciousness. Hindu traditions, with their cyclical time scales and multiple manifestations of divine consciousness throughout the cosmos, similarly offer frameworks potentially compatible with diverse forms of extraterrestrial intelligence.

Indigenous religious traditions worldwide often already include star beings or sky people in their cosmologies, recognizing spiritual connections between Earth and the cosmos. Many of these traditions emphasize relationships and interconnections between different classes of beings rather than placing humans in a position of cosmic privilege, potentially offering more integrative frameworks for conceptualizing alien intelligence.

Yet despite these adaptations and potential compatibilities, religious frameworks still face a fundamental challenge when confronting the possibility of truly alien intelligence: they remain products of human experience and cognition, utilizing conceptual categories evolved on Earth. Even the most expansive religious cosmologies tend to extend recognizably human qualities—consciousness, morality, purpose, language—into non-human domains rather than embracing forms of existence that might operate according to entirely different principles.

This anthropocentrism isn't unique to religion; as previous chapters have argued, it pervades scientific extraterrestrialist thinking as well. But religious traditions face a particular tension because many explicitly claim to offer comprehensive cosmic frameworks revealed by transcendent sources. If these frameworks still center human experience despite their divine origin, this raises questions about whether even divinely-inspired human understanding can transcend the cognitive limitations of our evolved psychology and terrestrial experience.

Exotheology and Its Limits

IN RESPONSE TO THESE challenges, the field of "exotheology" (sometimes called "astrotheology") has emerged as an attempt to systematically address the theological implications of extraterrestrial intelligence. Pioneered by scholars like Ted Peters, Michael Waltemathe, and David Wilkinson, exotheology examines how religious doctrines might accommodate the discovery of alien life and how such a discovery might transform religious understanding.

Exotheology represents a serious intellectual attempt to prepare religious frameworks for potential contact with extraterrestrial intelligence. It asks important questions: Would contact with aliens trigger religious crises? How might religious communities interpret alien religious beliefs or lack thereof? Could comparative theology extend beyond Earth to include alien spiritual perspectives? These inquiries demonstrate theological willingness to engage with the profound implications of a potentially wider cosmic community.

However, exotheology also reveals the limitations of religious approaches to extraterrestrial intelligence. Most exotheological exploration remains firmly anchored in human theological categories. It typically assumes that aliens would have recognizable "religious" or "spiritual" concepts analogous to human traditions, that they would engage in activities recognizable as "worship" or "moral reasoning," and that their relationship with ultimate reality could be understood through concepts like "God," "salvation," or "enlightenment."

This conceptual limitation becomes clear when exotheologians attempt to imagine alien religious perspectives. The scenarios proposed typically resemble human religious diversity projected onto non-human beings—aliens as monotheists, polytheists, pantheists, or atheists. Even when exotheologians attempt radical speculation, they generally recombine elements of human religious history rather than conceptualizing truly alien modes of relating to ultimate reality.

More fundamentally, exotheology rarely questions whether the category of "religion" itself might be a specifically human phenomenon rather than a universal feature of intelligent life. Religion emerges from distinctly human cognitive capacities (abstract symbolism, theory of mind, mortality awareness), evolutionary pressures (social cohesion, existential anxiety management), and historical circumstances. Aliens with different cognitive architectures, evolutionary histories, and social structures might organize their relationship to ultimate questions in ways that don't map onto human religious categories at all.

This isn't to dismiss exotheology's value. By grappling with how religious frameworks might accommodate alien intelligence, exotheologians perform important intellectual work within their traditions. They demonstrate religion's capacity for growth and adaptation in response to new knowledge. But exotheology, like all human theological enterprises, cannot escape the anthropocentric limitations of human cognition and experience. It can stretch religious categories to accommodate theoretical aliens, but it cannot truly think beyond the human conceptual frameworks from which religion emerges.

Ancient Astronauts: Pseudo-Religious Mythology

WHILE MAINSTREAM RELIGIONS have cautiously incorporated the possibility of extraterrestrial life into their theological frameworks, the ancient astronaut theory has developed into what might be called a pseudo-religious mythology of alien intervention. Popularized by authors like Erich von Däniken (*Chariots of the Gods?*, 1968) and Zecharia Sitchin (*The 12th Planet*, 1976), this theory reinterprets ancient religious texts, archaeological sites, and artwork as evidence of prehistoric extraterrestrial contact.

According to ancient astronaut theorists, aliens visited Earth in antiquity, influenced human evolution, taught advanced technologies to early civilizations, and were worshipped as gods by "primitive" humans who misunderstood their technology. The massive stones of Baalbek, the precision of the Great Pyramid, the enigmatic Nazca lines—all are reinterpreted as evidence of alien technological intervention rather than human achievement.

This theory functions as a modern creation mythology, addressing contemporary existential questions about human origins and cosmic significance through narratives of alien intervention. It offers a technological reframing of traditional religious origin stories, replacing divine creation with extraterrestrial genetic engineering and divine revelation with alien technological instruction. Like traditional religious mythologies, it provides narrative coherence to human existence and places humanity within a larger cosmic context.

The ancient astronaut theory also performs other traditionally religious functions. It offers hermeneutic frameworks for interpreting ancient texts and artifacts, creates communities of believers who share specialized knowledge, provides explanatory narratives for otherwise mysterious phenomena, and promises future revelations or returns of the alien "gods." Some variants even incorporate apocalyptic elements, with predictions of alien return coinciding with global transformation.

Despite these religious characteristics, ancient astronaut theories typically present themselves as scientific or historical hypotheses rather than belief systems. They use scientific terminology, appeal to physical evidence, and position themselves as rational alternatives to traditional religious accounts. This science-mimicking quality allows the ancient astronaut mythology to flourish in a secular age, offering religious-like meaning within a supposedly empirical framework.

However, the ancient astronaut theory ultimately resolves neither scientific nor religious questions about extraterrestrial intelligence. Scientifically, it fails basic standards of evidence, relying on selective interpretation of ambiguous artifacts, misunderstanding of ancient construction techniques, and decontextualized analysis of religious imagery. Religiously, it reduces the spiritual and symbolic dimensions of ancient traditions to literal misunderstandings of technology, failing to honor these traditions on their own terms.

Most significantly for our purposes, the ancient astronaut theory perpetuates rather than challenges anthropocentrism in extraterrestrialist thinking. Despite featuring non-human beings, these narratives place humanity at the center of alien concern, imagining extraterrestrials traversing interstellar distances specifically to influence human development. They project recognizably human motivations onto aliens—curiosity, benevolence, exploitation, control—rather than imagining truly alien psychologies or purposes.

The ancient astronaut mythology thus represents neither scientific advancement nor genuine religious innovation in conceptualizing extraterrestrial intelligence. Instead, it recombines elements of both domains within a fundamentally anthropocentric framework, creating a pseudo-religious narrative that tells us more about contemporary human anxieties and aspirations than about the possibility of alien existence.

The Religious Impulse in Scientific Cosmologies

WHILE RELIGIOUS FRAMEWORKS explicitly incorporate supernatural elements into their cosmologies, ostensibly secular scientific accounts of the cosmos often contain subtle religious or quasi-religious dimensions. These elements reveal the persistence of religious impulses even within frameworks that reject traditional religious authority, suggesting that neither science nor religion has fully resolved humanity's struggle to place itself within a potentially inhabited cosmos.

Consider how contemporary scientific cosmology often employs language and narratives with religious resonances. The Big Bang is described as the "creation" of the universe. Discussions of fine-tuning and anthropic principles suggest a universe somehow oriented toward producing intelligent observers. The search for extraterrestrial intelligence is framed as a quest to discover whether we are "alone" in the universe—a question with existential and spiritual dimensions as much as scientific ones.

The concept of the "Rare Earth" hypothesis, which argues that complex life may be extremely uncommon in the universe due to the improbable combination of factors that enabled it on Earth, sometimes takes on tones reminiscent of divine election or special creation. Similarly, the "simulation hypothesis" proposed by philosopher Nick Bostrom, suggesting that our reality might be a computer simulation created by an advanced civilization, bears structural similarities to creationist cosmologies, with the simulator playing a god-like role.

Even the Drake Equation, discussed in Chapter 3, serves a quasi-religious function in extraterrestrialist discourse. Despite its scientific framing, it operates as what religious studies scholars might call a "ritual formula"—a symbolic mathematical expression that organizes communal thinking about cosmic questions rather than producing empirically verifiable predictions. The ritualistic recitation of the Drake Equation in discussions of extraterrestrial intelligence resembles how religious communities use creeds or prayers to structure their cosmic understanding.

These religious dimensions of scientific cosmology aren't inherently problematic. They reflect the deep human need to find meaning and orientation within the cosmos, a need that transcends the institutional boundaries between science and religion. However, they do suggest that scientific frameworks, like religious ones, remain influenced by human psychological needs and cultural patterns rather than providing purely objective perspectives on cosmic questions.

Perhaps most significantly, these religious impulses within scientific cosmology often reinforce anthropocentrism in extraterrestrialist thinking. The concern with whether we are "alone," the focus on finding life "like us," the anxiety about our cosmic significance—these preoccupations reflect fundamentally human existential concerns projected onto the universe at large. They reveal how even ostensibly objective scientific inquiry remains shaped by subjective human needs for meaning, connection, and cosmic purpose.

This persistent religious impulse within scientific cosmology suggests that the science/religion divide may be less significant for extraterrestrialist thinking than the anthropocentric/ post-anthropocentric divide. Both traditional religious frameworks and ostensibly secular scientific ones typically center human perspectives and concerns when conceptualizing extraterrestrial possibilities. Both struggle to imagine genuinely alien modes of existence that might operate according to principles beyond human comprehension.

Spiritual Perspectives Beyond the Human/ Alien Binary

DESPITE THE LIMITATIONS of conventional religious approaches to extraterrestrial intelligence, certain spiritual perspectives offer resources for moving beyond the human/alien binary that constrains much extraterrestrialist thinking. These perspectives don't necessarily resolve the fundamental epistemological challenges of conceptualizing truly alien intelligence, but they do suggest pathways for expanding our cosmic imagination beyond conventional anthropocentric categories.

Mystical traditions across religious boundaries often emphasize direct experiential knowledge that transcends ordinary cognitive categories. The apophatic or "negative" theological traditions—represented in Christianity by figures like Pseudo-Dionysius and Meister Eckhart, in Judaism by aspects of Kabbalah, in Islam by certain Sufi practices, and in Hinduism by the neti neti ("not this, not that") approach of Advaita Vedanta—all suggest that ultimate reality transcends human conceptual frameworks. These traditions develop practices and perspectives for approaching what lies beyond human categories, potentially offering resources for engaging with truly alien modes of existence.

Indigenous spiritual traditions often emphasize radical interconnection between all entities and recognize agency and personhood in what Western frameworks categorize as "inanimate" nature. These perspectives challenge the sharp boundaries between conscious and non-conscious, living and non-living, that characterize much Western extraterrestrialist thinking. By recognizing consciousness and agency distributed throughout the natural world rather than limited to human-like entities, these traditions suggest ways of recognizing alien intelligence that might not conform to conventional Western criteria.

Contemporary nature-based spiritualities, including some expressions of deep ecology, bioregionalism, and Gaian spirituality, extend consciousness and agency to planetary systems themselves. The Gaia hypothesis, proposed by scientist James Lovelock and developed in more spiritual directions by figures like Stephan Harding and David Abram, suggests that Earth's biosphere functions as a self-regulating system with emergent properties that could be considered a form of planetary intelligence. Such perspectives offer frameworks for conceptualizing alien intelligence operating at scales and through processes quite different from individual organisms.

Transpersonal psychology and consciousness studies explore states of awareness that transcend ordinary individual consciousness. Developed by figures like Stanislav Grof, Ken Wilber, and Ralph Metzner, these approaches investigate non-ordinary consciousness states—including those induced by meditation, breathwork, or psychedelics—as potential windows into forms of awareness beyond conventional human parameters. While controversial within mainstream psychology, these explorations suggest that human consciousness itself may contain potentials for experiencing reality in ways that transcend ordinary anthropocentric limitations.

These diverse spiritual perspectives don't provide ready-made solutions to the challenges of conceptualizing extraterrestrial intelligence. Each emerges from specific human cultural contexts and remains subject to human cognitive constraints. However, they do suggest that human spirituality contains resources for approaching the truly other—for developing practices, perspectives, and experiences that push against the boundaries of conventional anthropocentric thinking.

Moreover, these perspectives challenge the common extraterrestrialist assumption that advanced intelligence naturally develops along a linear pathway from spirituality to scientific materialism. They suggest instead that genuine cognitive advancement might involve integrating rather than transcending spiritual modes of knowing—developing more sophisticated approaches to consciousness, interconnection, and transpersonal reality rather than abandoning these dimensions in favor of purely materialist frameworks.

This integration might characterize some forms of extraterrestrial intelligence, particularly those that have evolved far beyond current human development. Rather than projecting either conventional religious frameworks or scientific materialism onto hypothetical aliens, we might consider the possibility of intelligences that have developed more integrated approaches to physical and transpersonal dimensions of reality—approaches that might appear either incomprehensibly mystical or advanced beyond recognition as "spiritual" from our limited perspective.

Critique of Both Scientific Materialism and Religious Anthropocentrism

THE LIMITATIONS OF both conventional religious frameworks and scientific materialism in conceptualizing extraterrestrial intelligence point toward a more fundamental critique. Both approaches typically fail to transcend anthropocentrism not because of their specific content but because of their epistemological structures—the ways they organize knowledge and establish boundaries between knowing subjects and known objects.

Scientific materialism, despite its claims to objectivity, remains constrained by human sensory systems, cognitive architecture, and technological extensions of these faculties. It privileges certain forms of knowledge—quantifiable, reproducible, technologically detectable—while marginalizing others—subjective, qualitative, directly experienced. This epistemological framework inevitably filters potential extraterrestrial realities through human perceptual and conceptual screens, potentially rendering some forms of alien existence literally imperceptible or incomprehensible within scientific parameters.

Religious frameworks, despite their claims to revealed knowledge transcending human limitations, remain products of human psychology, language, and cultural evolution. They organize cosmic reality according to categories meaningful within human experience—personhood, agency, morality, purpose—and typically position humanity in special relationship with divine reality. Even the most expansive religious cosmologies tend to extend human-derived categories throughout the cosmos rather than genuinely transcending anthropocentric perspectives.

Both frameworks typically maintain sharp boundaries between knowing subjects and known objects, between self and other, that may constrain our ability to conceptualize genuinely alien modes of existence. Both assume that knowledge about extraterrestrial intelligence would come through established channels—either scientific detection or religious revelation—rather than requiring radical transformations in how humans know and relate to cosmic others.

A more adequate approach might require what philosopher of science Bruno Latour calls "compositionism"—developing frameworks that recognize both human and non-human agencies as co-creators of shared realities rather than positioning humans as either neutral observers or special recipients of divine attention. Such an approach would acknowledge how human knowledge emerges through specific material practices, technological mediations, and embodied experiences rather than claiming unmediated access to objective reality or divine truth.

This critique doesn't suggest abandoning either scientific or religious approaches to extraterrestrial possibilities. Each offers valuable resources for expanding human cosmic understanding. Science provides methodological rigor, empirical grounding, and technological extensions of human perception. Religion offers experiential depth, ethical frameworks, and recognition of dimensions beyond material causality. Both will remain essential aspects of humanity's engagement with cosmic questions.

Rather, this critique suggests the need for greater epistemological humility in both domains—recognition of how thoroughly human knowledge is shaped by evolutionary history, embodied experience, and cultural context. Such humility might open space for what philosopher Donna Haraway calls "staying with the trouble"—remaining in the uncomfortable recognition of our limitations rather than prematurely resolving them through either scientific certainty or religious assurance.

For extraterrestrialist thinking, this epistemological humility would mean acknowledging that our conceptions of alien intelligence, whether scientific or religious, tell us more about ourselves than about what might actually exist beyond Earth. It would mean recognizing that the very categories we use to conceptualize extraterrestrial possibilities—intelligence, consciousness, civilization, communication—emerge from specifically human experiences and may not adequately capture truly alien modes of existence.

Toward a Post-Anthropocentric Spiritual Cosmology

IF BOTH CONVENTIONAL religious frameworks and scientific materialism remain constrained by anthropocentrism in their approaches to extraterrestrial intelligence, what alternatives might be possible? Can we imagine a post-anthropocentric spiritual cosmology that might better prepare us for encountering—or at least conceptualizing—genuinely alien modes of existence?

Such a cosmology would likely include several key elements:

1. **Recognition of diverse forms of agency** beyond conventional human categories, acknowledging that consciousness, intelligence, and selfhood might manifest in ways radically different from human experience
2. **Epistemological pluralism** that values multiple ways of knowing—scientific, experiential, intuitive, embodied—without privileging any single approach as the exclusive path to understanding cosmic reality

3. **Attention to scale** in both temporal and spatial dimensions, developing frameworks that can conceptualize intelligence operating at timescales from microseconds to eons and at spatial scales from quantum to galactic
4. **Comfort with liminality and ambiguity**, embracing states between conventionally distinct categories (living/non-living, conscious/non-conscious, material/spiritual) rather than forcing phenomena into binary classifications
5. **Bioregional grounding** that acknowledges how all consciousness emerges from specific material contexts while remaining open to modes of existence adapted to environments radically different from Earth
6. **Multispecies ethics** extending moral consideration beyond humans to diverse forms of existence based on their intrinsic qualities rather than their similarity to humanity
7. **Linguistic and conceptual innovation** developing new vocabularies and frameworks for phenomena that exceed conventional human categories

Elements of such a post-anthropocentric spiritual cosmology already exist within various traditions. Some forms of process theology, developing from the work of Alfred North Whitehead and Charles Hartshorne, conceptualize divine reality as emerging through the experiences of all entities rather than imposing order from outside. Indigenous traditions worldwide recognize agency and personhood in non-human entities without requiring them to conform to human psychological patterns. Certain strands of mysticism emphasize direct encounter with otherness that transcends conventional categories.

Contemporary philosophical movements like posthumanism, new materialism, and speculative realism also offer resources for developing more expansive frameworks. Thinkers like Jane Bennett, Timothy Morton, and Bruno Latour have explored how agency, consciousness, and relationality might extend beyond conventional human categories, suggesting possibilities for recognizing truly alien forms of existence that neither religious nor scientific frameworks typically accommodate.

A post-anthropocentric spiritual cosmology wouldn't resolve all the challenges of conceptualizing extraterrestrial intelligence. Human knowledge inevitably emerges from human embodied experience and cognitive structures; we cannot simply step outside these parameters. But such a cosmology might create more space for acknowledging what we don't and perhaps can't know—for recognizing the limits of our cosmic understanding while remaining open to possibilities beyond our current conception.

This approach suggests a different orientation toward the search for extraterrestrial intelligence—less about confirming human cosmic uniqueness or companionship, more about opening ourselves to the possibility of encountering genuine otherness. It shifts the question from "Are we alone?" (a question centered on human existential concerns) to "What exists?" (a question that decenters human perspective in favor of cosmic realities that may or may not include entities resembling or concerning themselves with us).

Such a post-anthropocentric orientation doesn't abandon the human perspective—we cannot escape being human—but it does hold that perspective more lightly, recognizing it as one among many possible ways of experiencing and organizing cosmic reality. It suggests approaching potential extraterrestrial intelligence not as an answer to human existential questions but as an opportunity to expand human conception beyond its current anthropocentric limitations.

Conclusion: Beyond Cosmic Purpose

THIS CRITICAL EXAMINATION of religious approaches to extraterrestrial intelligence returns us to a central theme of this book: the anthropocentrism that pervades human cosmic thinking across domains. Neither science nor religion, despite their different methodologies and knowledge claims, has fully transcended the human-centered frameworks through which we inevitably conceptualize the cosmos and its potential inhabitants.

Religious traditions have offered expansive cosmologies incorporating non-human intelligences, from angels to devas to jinn. They have developed sophisticated theological responses to the possibility of extraterrestrial life. They have provided experiential practices for approaching realities beyond ordinary human categories. Yet they have typically extended recognizably human qualities—personhood, morality, purpose—throughout the cosmos rather than genuinely transcending anthropocentric perspectives.

Similarly, scientific approaches have expanded human perception through technological instruments, developed mathematical frameworks for calculating extraterrestrial probabilities, and proposed search strategies for detecting alien technosignatures. Yet they too have generally projected human-derived concepts—intelligence as problem-solving, civilization as energy utilization, communication as intentional signaling—onto hypothetical extraterrestrial entities.

Both scientific and religious approaches to extraterrestrial intelligence typically presuppose some form of cosmic purpose or meaning—whether divine plan or evolutionary progression—that makes intelligence comprehensible within larger frameworks. Both struggle with the possibility that intelligent life might be neither cosmically significant (as religion often suggests) nor cosmically common (as science often calculates), but rather a contingent phenomenon emerging in particular circumstances without larger meaning or purpose.

Perhaps the most profound challenge of extraterrestrialist thinking is not determining whether aliens exist or what they might be like, but confronting the possibility that their existence or non-existence might not matter in the ways we typically assume. Aliens, if they exist, might be neither the answer to humanity's cosmic loneliness nor the threat to human cosmic uniqueness, but simply other contingent phenomena emerging through their own evolutionary and cultural histories, indifferent to human existential concerns.

This possibility challenges both religious cosmic narratives centered on divine purpose and scientific narratives centered on evolutionary progression. It suggests that the universe might contain intelligent beings without containing cosmic meaning—that consciousness might emerge in various forms throughout the cosmos without validating human existential needs for significance or connection.

Such a perspective doesn't render the search for extraterrestrial intelligence pointless. Understanding what cosmic companions might exist remains a profound scientific question and a potentially transformative discovery. But it does suggest conducting this search with greater awareness of the human needs and projections that shape it—with recognition that what we find, if anything, may tell us more about ourselves than about the universe at large.

In the end, religion does not help resolve the fundamental questions of extraterrestrial existence not because religious perspectives lack value, but because these questions ultimately transcend both religious and scientific frameworks. They touch on the limits of human knowledge itself, on the boundaries between what we can know and what we can only speculate about from our inevitably anthropocentric perspective.

This recognition need not lead to nihilism or abandonment of the search. Rather, it might inspire what philosopher Isabelle Stengers calls "cosmic politics"—approaches to cosmic questions that acknowledge their entanglement with human social and psychological concerns without reducing them to these concerns. Such politics would neither dismiss extraterrestrial questions as mere projection nor treat them as separable from human contexts, but would engage them as sites where human and potentially non-human realities meet in complex, co-constitutive relationships.

Through such engagement, we might develop not final answers about extraterrestrial intelligence but more nuanced questions—questions that acknowledge both the limitations of human perspective and the genuine possibility of cosmic otherness beyond our current conception. These questions might prepare us, if not to comprehend truly alien intelligence, then at least to recognize the boundaries of our comprehension—a recognition that might itself constitute a form of cosmic wisdom.

Chapter 7: The Epistemological Limits of Alien Recognition

"If a lion could speak, we would not understand him." — Ludwig Wittgenstein

The Recognition Problem

In July 2015, Russian billionaire Yuri Milner announced the launch of Breakthrough Listen, a $100 million initiative to search for intelligent extraterrestrial life. The project would use some of the world's most powerful radio telescopes to scan the nearest million stars for artificial radio or laser signals that might indicate technological civilizations. Alongside this listening project, Milner also funded Breakthrough Message, a competition to design digital messages that might someday be sent to alien civilizations, and Breakthrough Starshot, an ambitious plan to send miniature probes to nearby star systems.

These projects represent perhaps the most well-funded effort in history to detect and potentially communicate with extraterrestrial intelligence. Yet they rest on a profound assumption that has received surprisingly little critical examination: that we would recognize alien intelligence if we encountered it, and that aliens would recognize us. This assumption—that intelligence is mutually recognizable across species boundaries—underlies virtually all SETI efforts and extraterrestrialist thinking. But is it justified?

This chapter argues that the recognition of truly alien intelligence may be fundamentally constrained by epistemological limits—boundaries to what and how we can know that emerge from our evolved cognitive architecture, our embodied experience as specific biological organisms, and our embeddedness in human cultural frameworks. These constraints don't make alien recognition absolutely impossible, but they do suggest that our capacity to recognize intelligence unlike our own may be far more limited than conventional extraterrestrialist thinking acknowledges.

The recognition problem has two symmetric aspects: our ability to recognize alien intelligence and aliens' potential ability to recognize us. Both directions of recognition face similar constraints emerging from the species-specific nature of cognition and communication. These constraints create what might be called "recognition asymmetries"—gaps between what exists and what can be recognized as existing by differently embodied and evolved minds.

Understanding these epistemological limits doesn't render the search for extraterrestrial intelligence futile. Rather, it invites more sophisticated approaches—methods that acknowledge cognitive constraints while developing practices that might partially transcend them. By recognizing the boundaries of recognition itself, we might develop more nuanced searches for cosmic companions that don't simply project human patterns onto the cosmos but remain genuinely open to otherness beyond our current conception.

Signal Interpretation and the Problem of Mutual Comprehension

THE CLASSIC APPROACH to SETI focuses on detecting electromagnetic signals—radio transmissions or laser pulses—that appear artificial rather than natural in origin. This approach assumes not only that advanced civilizations would use electromagnetic communication but also that we could distinguish meaningful signals from cosmic noise and potentially interpret their content.

This assumption faces profound challenges even before considering truly alien modes of cognition. Consider the difficulties in a relatively simple case: detecting and interpreting a deliberate mathematical message sent by humanlike aliens using radio waves. Even this "best-case scenario" reveals multiple layers of interpretation problems:

1. **Distinguishing signal from noise**: Identifying artificial patterns amid the complex background of natural cosmic radio emissions
2. **Identifying intended communication**: Differentiating between deliberate messages and technological byproducts (like radar or broadcast leakage)
3. **Decoding physical parameters**: Determining modulation methods, frequencies, timing sequences
4. **Extracting syntax and structure**: Recognizing patterns that constitute "grammar" or organization
5. **Inferring semantics**: Moving from pattern recognition to meaning attribution
6. **Contextualizing content**: Placing meanings within conceptual frameworks that make them comprehensible

Each of these interpretive layers introduces possibilities for misunderstanding or non-recognition. A signal might be powerful enough to detect but structured in ways we don't recognize as artificial. It might use modulation techniques we haven't developed. Its syntax might follow logical principles unfamiliar to human mathematics or linguistics. Its semantic content might reference concepts, entities, or relationships that have no analogs in human experience.

These challenges become exponentially more complex when we consider truly alien intelligences rather than humanlike ones. Aliens with different sensory modalities, cognitive architectures, or evolutionary histories might conceptualize reality in ways fundamentally different from human frameworks. Their mathematics might not prioritize counting or set theory but develop from entirely different foundational intuitions. Their "language" (if they use anything we would recognize as such) might organize reality according to categories that don't map onto human distinctions between objects, actions, and qualities.

Several approaches have been proposed to overcome these mutual comprehension challenges:

1. **Mathematical universality**: Using prime numbers, physical constants, or basic arithmetic as foundation points for communication
2. **Physical references**: Basing messages on shared physical realities like atomic spectra or fundamental particles
3. **Pictorial communication**: Transmitting images that might transcend linguistic differences
4. **Multi-level encoding**: Embedding keys within messages that teach recipients how to decode subsequent, more complex information

Each approach has merits, but all ultimately rely on significant shared cognitive frameworks. Mathematics as humans practice it emerges from specific cognitive capacities—abilities to distinguish discrete objects, recognize symmetries, generalize patterns—that are products of human evolution. While aliens might develop similar capacities convergently, they might also develop entirely different approaches to systematizing reality that wouldn't read as "mathematics" to human observers.

Even physical references assume shared frameworks for organizing perceptual experience. The distinction between elements in the periodic table, for instance, reflects not just objective natural patterns but specific human ways of categorizing and systematizing matter. Aliens might divide the physical world along entirely different lines—perhaps prioritizing process over substance, fields over particles, or relationships over entities in ways that would make even basic physical references mutually unintelligible.

These challenges suggest that mutual comprehension between radically different intelligences may be far more difficult than the optimistic assumptions of traditional SETI suggest. Communication might require not just decoding signals but developing entirely new conceptual frameworks—a process more akin to profound cross-cultural anthropology than to cryptography, and one with uncertain prospects for success even between the most open-minded parties.

The Impossibility of Recognizing Truly Alien Intelligence

BEYOND THE CHALLENGES of signal interpretation lies a deeper question: Would we recognize truly alien intelligence at all? Our conception of "intelligence" itself emerges from human cognitive processes and cultural frameworks. It reflects what human brains do and value rather than some objective, universal category.

Consider how human criteria for intelligence typically include:

1. **Problem-solving ability**: Finding novel solutions to environmental or conceptual challenges
2. **Pattern recognition**: Identifying regularities and anomalies in complex data

3. **Abstract reasoning**: Manipulating symbolic representations detached from immediate perception
4. **Adaptive learning**: Modifying behavior based on experience
5. **Intentionality**: Having purpose or goal-directedness
6. **Self-awareness**: Possessing a reflective sense of one's own existence
7. **Social cognition**: Understanding and predicting others' mental states

These criteria reflect capacities that evolved in humans to address specific challenges in our evolutionary environment. They're not arbitrary—they enabled our ancestors to survive and reproduce in particular ecological and social contexts—but they're also not universal necessities for complex organization or effective environmental interaction.

Intelligence that evolved under radically different conditions might prioritize entirely different cognitive capacities. Aliens adapted to environments with different physical properties, different resource distributions, or different competitive pressures might develop forms of environmental manipulation, information processing, or social coordination that bear little resemblance to human intelligence yet function effectively within their contexts.

Some speculative examples of such alternative intelligences might include:

1. **Distributed processing systems** without centralized awareness, more like slime molds or ant colonies than individual human minds
2. **Quantum cognition** operating through superposition and entanglement rather than classical computation

3. **Chemical intelligence** processing information through molecular interactions rather than electrochemical signals
4. **Temporal intelligence** operating at time scales vastly slower or faster than human cognition
5. **Field-based awareness** emerging from electromagnetic or gravitational interactions rather than discrete neural networks

Such alternative intelligences might not engage in behaviors we typically associate with intelligence. They might not build tools, create symbolic representations, communicate through discrete messages, solve problems through logical analysis, or demonstrate self-awareness in recognizable forms. Their "technology" might be indistinguishable from biological processes or natural phenomena. Their "communication" might occur through media or methods we don't monitor or recognize as communicative.

The concept of "Humblebacteria," proposed by mathematician and science fiction author Vernor Vinge, illustrates this recognition problem. Vinge imagined a bacterial species with distributed intelligence far exceeding human capabilities but manifesting in ways humans would never recognize as intelligent—through chemical exchanges, subtle environmental modifications, and collective adaptations that would appear as natural processes rather than as products of conscious intelligence.

The Humblebacteria thought experiment suggests that intelligence might be literally under our feet (or in the clouds, or between the stars) without humans having any way to recognize it as such. Our scientific instruments might detect its effects but categorize them as ordinary natural phenomena. Our SETI searches might scan right past its communications, failing to distinguish them from background processes. Even direct contact might result in mutual non-recognition, with neither humanity nor the alien intelligence registering the other as a meaningful entity.

This possibility—that intelligence might be fundamentally non-recognizable across species boundaries—challenges the core assumption of traditional extraterrestrialist thinking. If intelligence is recognizable only within relatively narrow parameters of similarity, then our search for extraterrestrial intelligence may be not just practically difficult but conceptually limited to finding entities somewhat like ourselves—a tiny subset of possible cosmic intelligences.

Wittgenstein's Language Games and the Limits of Cross-Species Communication

THE PHILOSOPHICAL FOUNDATIONS for understanding the limits of cross-species communication were powerfully articulated by Ludwig Wittgenstein in his later work, particularly in *Philosophical Investigations* (1953). Wittgenstein's insights into the nature of language and meaning suggest fundamental constraints on communication between differently embodied and socially situated beings—constraints that apply with particular force to hypothetical communication with extraterrestrial intelligence.

Wittgenstein argued that language gains meaning not through direct correspondence with external reality but through its use within what he called "language games"—rule-governed social practices embedded in particular "forms of life." The meaning of words emerges from how they function within these shared practices rather than from pointing to objective, language-independent realities. His famous dictum "meaning is use" encapsulates this pragmatic approach to language.

This perspective suggests that communication requires not just shared symbols but shared contexts of use—common practical activities, social structures, and physical environments that give language its functional significance. When Wittgenstein wrote, "If a lion could speak, we would not understand him," he wasn't merely noting a translation problem but identifying a deeper issue: the lion's language would be embedded in lion experiences, lion social practices, and lion phenomenology that humans don't share.

Applied to potential extraterrestrial communication, this Wittgensteinian insight suggests profound limitations. Aliens would have evolved in environments potentially radically different from Earth, with different perceptual systems, different social structures, and different practical challenges shaping their cognition and communication. Their "language games" would be embedded in forms of life that humans have never experienced and potentially cannot experience due to our different biological embodiment.

Consider attempts to communicate basic spatial concepts. Human spatial cognition is shaped by our specific embodiment—our bilateral symmetry, upright posture, forward-facing eyes, and habitation on a planetary surface with consistent gravity. These conditions create the foundations for concepts like "up/down," "left/right," "front/back" that seem objective to us but actually reflect our particular embodied perspective.

Aliens with radically different embodiment—perhaps radially symmetric beings, or entities inhabiting gas giants without solid surfaces, or organisms experiencing four-dimensional spatial perception—would develop spatial concepts organized according to entirely different parameters. Their "language games" for discussing space would embed assumptions, distinctions, and relationships that might be literally unthinkable within human cognitive frameworks.

Similar challenges would apply across all conceptual domains. Time, causality, identity, relationship—all concepts that seem universal from within human experience—emerge from specific interactions between human biology, human sociality, and Earth's particular environmental conditions. Alien concepts in these domains might be organized according to principles that don't map onto human distinctions, creating fundamental barriers to mutual comprehension.

These Wittgensteinian insights don't necessarily imply that all communication with extraterrestrial intelligence would be impossible. They do suggest, however, that such communication would require more than deciphering alien symbols or translating alien grammar. It would require developing new conceptual frameworks—new "language games"—that could serve as bridges between radically different forms of life. Such bridging frameworks might need to be constructed cooperatively over extended periods of interaction, with no guarantee of comprehensive mutual understanding.

The alternative—assuming that mathematics, logic, or physics provide universal languages that transcend biological and cultural differences—likely underestimates how thoroughly these domains themselves are shaped by human cognitive architecture and embodied experience. What appears as universal truth from within human perspective may reveal itself as species-specific cognition when encountered from radically different viewpoints.

Lessons from Terrestrial Interspecies Communication

WHILE EXTRATERRESTRIAL intelligence remains hypothetical, Earth provides numerous examples of non-human intelligence that might offer insights into the challenges of cross-species communication. Decades of research with great apes, cetaceans, corvids, cephalopods, and other intelligent animals have revealed both possibilities and limitations in interspecies understanding that may inform extraterrestrialist thinking.

Studies teaching symbolic language to great apes offer particularly relevant lessons. Beginning with the pioneering work of researchers like Allen and Beatrix Gardner with the chimpanzee Washoe in the 1960s, numerous projects have taught modified American Sign Language or other symbolic systems to chimpanzees, bonobos, gorillas, and orangutans. These studies demonstrated that apes can acquire substantial vocabularies, combine symbols in rule-governed ways, and use signs communicatively.

However, ape language research also revealed significant limitations. Great apes rarely generate novel grammatical combinations, seldom ask questions, and show limited grasp of syntactic relationships that human children master easily. Their use of symbols remains primarily focused on immediate needs and concrete situations rather than abstract concepts or hypothetical scenarios. Most significantly, apes raised with human language don't spontaneously teach this language to their offspring—it remains a learned interface with humans rather than becoming integrated into ape social life.

These limitations persist despite great apes sharing approximately 98% of their genetic material with humans, having evolved in the same planetary environment, and possessing sensory systems and social structures broadly similar to human ones. The communication gap between humans and their closest evolutionary relatives—despite extensive shared biology and researchers' extraordinary efforts to bridge this gap—suggests the potential magnitude of communication challenges with truly alien intelligence.

Research with marine mammals reveals different aspects of interspecies communication challenges. Dolphins and whales possess complex communication systems and demonstrate sophisticated cognitive abilities, including apparent self-awareness, social learning, and tool use. Yet despite decades of research, scientists have made limited progress in deciphering cetacean communication or establishing two-way information exchange beyond simple trained responses.

The cetacean case illustrates how different environmental adaptations shape communication systems. Underwater acoustic environments favor different signal properties than terrestrial ones. Dolphins' three-dimensional social space creates different communicative contexts than human face-to-face interaction. Their primarily acoustic world (compared to humans' primarily visual one) organizes information according to different principles. Despite sharing the same planet and mammalian heritage, humans and cetaceans have developed communication systems with limited mutual intelligibility.

Studies of corvid cognition (ravens, crows, and their relatives) demonstrate yet another challenge: recognizing intelligence that operates through different mechanisms than human cognition. Corvids perform complex cognitive tasks—tool manufacture, causal reasoning, mental time travel, theory of mind—with brain structures radically different from the mammalian neocortex. Their intelligence emerges from neural architecture that evolved independently, demonstrating that similar cognitive capabilities can arise through different biological structures—a form of evolutionary convergence.

This convergent evolution of intelligence on Earth highlights both possibilities and challenges for extraterrestrial communication. It suggests that certain cognitive capabilities might evolve repeatedly under similar selective pressures, potentially creating some common ground. However, it also demonstrates that similar capabilities can emerge through different mechanisms, potentially organizing information in ways that don't map neatly across species boundaries despite functional similarities.

Perhaps the most profound lesson from terrestrial interspecies communication comes from considering intelligences we've failed to recognize. Western scientific tradition historically denied intelligence, consciousness, and complex emotional lives to many species now known to possess these qualities. Indigenous knowledge systems often recognized these capacities long before scientific validation—suggesting that cultural frameworks profoundly influence which intelligences we recognize and which we overlook.

This historical pattern of non-recognition—of intelligence hiding in plain sight—raises troubling questions for extraterrestrialist thinking. If we failed to recognize intelligence in fellow Earth species with whom we share evolutionary history and planetary context, how much more likely might we be to miss intelligence that evolved in radically different environments according to different principles? Our track record suggests a persistent bias toward recognizing only intelligence that sufficiently resembles our own—a bias that may significantly constrain our capacity to detect truly alien minds.

The Mirror Test: Would We Recognize Intelligence Unlike Our Own?

THE "MIRROR TEST," developed by psychologist Gordon Gallup Jr. in 1970, evaluates whether an animal recognizes itself in a mirror, indicating some form of self-awareness. Researchers place a mark on an animal where it can only be seen in a mirror and observe whether the animal uses the mirror to investigate the mark, demonstrating an understanding that the mirror image represents itself rather than another individual.

Only a few species consistently pass this test: great apes, elephants, dolphins, orcas, the Eurasian magpie, and some cleaner wrasses. This limited club is often cited as evidence for restricted distribution of self-awareness in the animal kingdom. However, the test itself embeds significant anthropocentric assumptions: it requires visual orientation, interest in physical appearance, sufficient motor control to investigate marks, and cognitive mapping between mirror images and bodily sensation—all capacities that might be irrelevant to self-awareness in differently embodied species.

A dog, with primary sensory orientation through smell rather than sight, might fail the mirror test while demonstrating self-awareness through olfactory discrimination. A bat, navigating primarily through echolocation, might conceptualize selfhood through acoustic rather than visual feedback. The mirror test reveals not just which animals possess self-awareness but which possess self-awareness structured similarly to human visual self-recognition—a much narrower category.

This mirror test limitation serves as a powerful metaphor for the broader challenges of alien intelligence recognition. Just as the mirror test evaluates only one culturally and biologically specific manifestation of self-awareness, our search for extraterrestrial intelligence may be testing for only specific manifestations of intelligence that reflect human cognitive patterns rather than universal markers of complex information processing.

Consider how current SETI approaches prioritize certain "intelligence markers":

1. **Narrowband radio transmissions** - Reflecting human communication technology rather than potentially numerous alternative communication methods
2. **Mathematical patterns** - Privileging human-style abstract symbolic reasoning over alternative information-organization systems
3. **Technosignatures like megastructures** - Assuming intelligence manifests through large-scale environmental manipulation rather than alternative adaptations
4. **Atmospheric modifications** - Looking for industrial byproducts similar to human technological development rather than alternative civilization patterns

Each approach functions as a kind of "mirror test" that would identify intelligences similar to humanity while potentially missing vast ranges of alternative intelligence. Civilizations that communicate through gravitational manipulation, coordinate through chemical exchanges, process information through quantum entanglement, or structure their activities according to principles entirely different from human societies might be invisible to these detection methods.

More fundamentally, our conception of what constitutes evidence of intelligence reflects human cognitive biases. We recognize as intelligent those processes that resemble human thought: sequential causal reasoning, symbolic representation, intentional design. Processes that organize complexity through different mechanisms—emergent self-organization, distributed processing, non-linear causality—may not register as intelligence despite potentially equal or greater computational sophistication.

This recognition bias appears even in how we conceptualize alien technological development. The Kardashev Scale, discussed in Chapter 4, measures advancement through energy utilization—a metric reflecting industrial capitalist values rather than universal developmental principles. A civilization might develop sophisticated technologies focused on information density rather than energy consumption, sustainability rather than expansion, or cognitive enhancement rather than physical manipulation—all pathways that might appear "primitive" according to Kardashev metrics despite potentially representing more advanced approaches to intelligence.

The mirror test metaphor suggests that SETI might benefit from developing multiple, complementary approaches based on different conceptions of intelligence—creating various "mirrors" that might reflect different manifestations of cosmic complexity. Rather than asking simply "Are they like us?" we might develop methods for detecting patterns that organize according to principles different from human cognition yet still represent complex information processing beyond random or simple physical processes.

Such an expanded approach would require extraordinary creativity in imagining alternative intelligences—a task inherently constrained by human cognitive architecture yet not entirely impossible. By leveraging insights from complex systems theory, non-human animal cognition, artificial intelligence research, and non-Western epistemologies, we might develop search methods that cast wider nets for cosmic complexity without assuming it must mirror human patterns.

Technological Signatures and Their Anthropocentric Bias

CURRENT APPROACHES to detecting extraterrestrial civilizations focus heavily on technological signatures—observable consequences of advanced technology that might be detectable across interstellar distances. These include:

1. **Artificial electromagnetic signals** - Radio broadcasts, radar, deliberate messaging
2. **Stellar modification** - Dyson spheres, stellar lifting, or other large-scale energy capture systems
3. **Atmospheric technosignatures** - Industrial pollutants, artificial chemicals, or other atmospheric modifications
4. **Planetary surface alterations** - Artificial lighting, large-scale agriculture, or urban heat signatures
5. **Interstellar travel indicators** - Propulsion signatures, artificial objects, or relativistic spacecraft effects

These search categories reflect a particular vision of technological development based on human history—one that prioritizes energy capture, resource exploitation, electromagnetic communication, and physical engineering. This vision assumes that all technological civilizations, regardless of their biology or cultural values, would develop along broadly similar trajectories, creating detectable signatures analogous to human industrial and post-industrial technology.

This assumption reveals a deep anthropocentrism in our technological imagination. It projects not just the fact of technology but specific forms of technological development based on terrestrial materials, terrestrial physics (as experienced at human scales), and human values and needs. It assumes that technologies resembling human approaches to energy, communication, and environmental modification represent universal or inevitable developmental pathways rather than contingent outcomes of Earth's specific conditions and human evolutionary history.

Alternative technological trajectories might produce entirely different signatures—or no detectable signatures at all from our current observational perspective. Consider some possibilities:

1. **Efficiency-focused technology** that minimizes energy use and environmental impact rather than maximizing power and scale
2. **Biochemical technology** that modifies and enhances biological processes rather than replacing them with mechanical alternatives
3. **Quantum manipulation** technologies operating at scales and through principles that produce few macroscopic observable effects
4. **Dimensional technologies** that utilize or access aspects of reality beyond standard four-dimensional spacetime
5. **Consciousness-enhancing technologies** that develop internal rather than external capacities

Such alternative technological trajectories might produce civilizations as or more advanced than human technology while generating minimal technosignatures detectable through current SETI methods. A civilization focused on perfect efficiency and recycling might produce almost no waste heat or pollution to detect. One that developed primarily through biological enhancement rather than mechanical engineering might create no artificial electromagnetic signals. Advanced quantum technologies might operate through principles that leave no observable macroscopic effects from our observational perspective.

Even more fundamentally, our conception of "technology" itself reflects specific human cognitive patterns and historical developments. We typically define technology as the deliberate manipulation of the environment through designed tools to achieve intended effects—a definition that centers human-style intentionality, causal reasoning, and subject-object distinction. This definition might not capture all systems for environmental modification, information processing, or problem-solving that could evolve in different cognitive architectures.

Some Earth examples suggest alternative conceptions: the agricultural systems of leafcutter ants, the architectural capabilities of termites, the molecular machinery of cellular metabolism. These complex functional systems achieve sophisticated environmental modifications without human-style intentional design. Similarly, alien "technology" might emerge through processes that don't map neatly onto human conceptions of engineering or invention—perhaps through directed evolution, emergent self-organization, or symbiotic development rather than discrete invention events.

This anthropocentrism in technological conception creates a potential blindspot in SETI: we search for civilizations whose technology resembles human approaches while potentially missing those that have developed along entirely different trajectories. Our search methods privilege loud, energetically intensive, expansionist technological civilizations over quiet, efficient, steady-state ones—a bias that may significantly skew our ability to detect cosmic intelligence.

A more comprehensive approach would require developing search methods calibrated to detect subtle signatures of alternative technological trajectories—perhaps minuscule but precise perturbations in natural systems, unusual chemical or quantum states, or patterns of biological modification that suggest deliberate enhancement. Such methods would require extraordinary creativity in imagining alternative technological trajectories and translating these imagined alternatives into concrete search parameters.

The Biosignature Problem: Recognizing Life That Doesn't Look Like Earth Life

BENEATH THE CHALLENGE of recognizing alien intelligence lies an even more fundamental question: Would we recognize alien life at all? Our conception of "life" itself emerges from Earth's specific biochemistry and evolutionary history. It reflects what terrestrial organisms do rather than universal principles of living systems.

When astrobiologists search for biosignatures—indicators of life on other worlds—they typically look for:

1. **Liquid water** - Essential to all known Earth life
2. **Carbon-based organic chemistry** - The foundation of terrestrial biochemistry
3. **Atmospheric disequilibrium** - Particularly oxygen and methane coexisting despite chemical instability
4. **Surface reflectance patterns** consistent with photosynthetic organisms
5. **Seasonal variations** suggesting biological cycles

These search parameters make sense given our single example of life, but they embed significant terrestrial bias. They assume that alien life would use similar chemical foundations, metabolic processes, and ecological patterns to Earth organisms. This assumption may be justified for life that emerged in broadly Earth-like conditions but becomes increasingly questionable for environments with dramatically different parameters.

Alternative biochemistries might include:

1. **Silicon-based** rather than carbon-based molecular foundations
2. **Ammonia or methane solvents** rather than water
3. **Different chirality** (mirror-image molecular orientation) from Earth life
4. **Alternative energy-capturing mechanisms** beyond photosynthesis or chemosynthesis as we know them
5. **Different genetic encoding systems** not based on DNA/RNA

Such alternative biochemistries might produce biosignatures unrecognizable through current detection methods or might create signatures that we would misinterpret as abiotic processes. For instance, silicon-based life might produce atmospheric signatures that resemble volcanic activity rather than biological processes from a terrestrial perspective. Life using radically different metabolic pathways might create atmospheric compositions with no analog on Earth, potentially leading us to overlook their biological origin.

Beyond alternative biochemistries lies an even more challenging possibility: forms of organization that function as life but do not share the defining characteristics of Earth life. Terrestrial biology defines life through criteria like:

- Metabolism (energy processing)
- Reproduction (self-replication)
- Homeostasis (internal regulation)
- Adaptation (environmental response)

- Organization (structural complexity)
- Growth and development

These criteria reflect the specific organizational patterns that evolved on Earth rather than universal requirements for complex self-sustaining systems. Alternative forms of organization might achieve similar functional goals through entirely different mechanisms—or might organize according to principles that don't map neatly onto terrestrial biological categories at all.

Speculative possibilities include:

1. **Plasma-based organizations** in stellar atmospheres, operating through electromagnetic principles rather than chemical ones
2. **Pattern-propagating systems** in interstellar dust clouds that maintain and replicate complex information without discrete "bodies"
3. **Quantum-coherent structures** that maintain organization through principles closer to physical fields than to chemical metabolism
4. **Geological-timescale systems** that process energy and information at rates thousands or millions of times slower than Earth life
5. **Symbiotic networks** without clear boundaries between "individuals" or "species" as we conceptualize them

Such alternative organizations might be literally imperceptible as "life" through current observational methods. Their biosignatures—if they produce any recognizable to us—might register as unusual but apparently abiotic physical phenomena rather than as indicators of living systems. Our existing categories for distinguishing between living and non-living systems might simply fail to capture their organizational principles.

This recognition problem becomes particularly acute because life and intelligence likely exist on a spectrum rather than representing discrete categories. Earth's history suggests a gradual emergence of complexity from geochemical cycles through primitive metabolism to complex multicellular organisms and eventually to intelligence. If alien life follows different organizational principles, the boundaries between abiotic processes, simple life, and complex intelligence might fall in entirely different places—or might not exist as meaningful distinctions at all.

These considerations suggest that our search for extraterrestrial life may be conceptually constrained in ways more fundamental than technical limitations of detection equipment. We may be not just limited in what we can see but in what we can recognize as significant once seen. Our categories for distinguishing life from non-life, intelligence from non-intelligence, technology from natural process—all emerge from Earth's specific history rather than representing universal distinctions applicable across all possible organizations of matter and energy.

Phenomenological Approaches to Alien Cognition

GIVEN THESE PROFOUND challenges in recognizing truly alien intelligence, are we left with nothing but radical skepticism about extraterrestrial cognition? Does the specificity of human perception and cognition create an insurmountable barrier to understanding minds unlike our own? Or might there be approaches that, while acknowledging these limitations, still allow for meaningful engagement with potential alien consciousness?

Phenomenology—the philosophical study of structure of experience and consciousness—offers potential resources for approaching these questions. Rather than assuming direct access to objective reality or retreating to pure subjectivity, phenomenology examines how consciousness constitutes its objects through intentional acts. This approach, developed by philosophers like Edmund Husserl, Martin Heidegger, and Maurice Merleau-Ponty, might provide conceptual tools for approaching alien cognition without imposing purely human categories.

Husserl's method of "phenomenological reduction" involves bracketing assumptions about the external world to examine the structures of consciousness itself. Applied to extraterrestrialist thinking, this method might help identify which aspects of our conception of intelligence reflect universal structures of consciousness and which reflect specifically human embodiment and cultural frameworks. While we cannot escape our human perspective entirely, we might develop greater awareness of its contingent elements.

Merleau-Ponty's emphasis on embodied cognition—the way consciousness emerges through bodily engagement with the world rather than existing as disembodied rationality—offers particular insight for conceptualizing alien intelligence. By recognizing how thoroughly human cognition is shaped by our specific bodies and sensory systems, we might imaginatively explore how differently embodied beings would constitute their worlds through alternative sensory-motor engagements.

These phenomenological approaches suggest several strategies for approaching alien cognition:

1. **Bracketing anthropocentrism**: Deliberately suspending assumptions about what constitutes intelligence,

consciousness, or meaningful communication to remain open to patterns that don't fit human categories
2. **Embodied imagination**: Using our understanding of how embodiment shapes cognition to speculatively explore how differently embodied beings might experience and conceptualize reality
3. **Attention to constitutive processes**: Focusing not just on the content of potential alien cognition but on the processes through which consciousness constitutes its objects, which might reveal structural similarities despite different content
4. **Seeking minimal common ground**: Identifying the most basic shared realities—existence within physical laws, temporal experience, environmental interaction—that might provide starting points for mutual comprehension

These approaches don't eliminate the epistemological challenges of alien recognition, but they offer more sophisticated frameworks than either naive assumptism (assuming aliens would think like humans) or complete skepticism (assuming alien cognition would be wholly incomprehensible). They suggest the possibility of partial, asymptotic understanding—approaching alien cognition without assuming complete comprehension.

Practical applications of phenomenological approaches might include:

1. **Multiple-paradigm SETI**: Developing search strategies based on different assumptions about how intelligence might manifest, consciously varying fundamental premises about cognition and technology
2. **Metacognitive protocols**: Creating guidelines for first-contact scenarios that incorporate awareness of human

cognitive biases and interpretation tendencies
3. **Embodiment variation**: Using virtual reality, sensory substitution, or other technologies to temporarily alter human perceptual frameworks, potentially expanding imagination about alternative cognitive modes
4. **Cognitive ethnography**: Studying the most radically different minds available on Earth—from cephalopods to artificial intelligence—to develop frameworks for approaching non-human cognition

These approaches share a common principle: using awareness of the limits of human understanding to create more receptive frameworks for potential engagement with non-human intelligence. Rather than assuming direct comprehension or surrendering to complete incommensurability, they cultivate what might be called "epistemological humility"—recognition of cognitive limitations coupled with commitment to reaching beyond them where possible.

Such epistemological humility might be essential not just for theoretical discussions of alien intelligence but for any potential actual contact. If humanity does someday encounter extraterrestrial intelligence, the greatest barrier to meaningful exchange might not be technological but conceptual—the challenge of recognizing patterns of meaning in systems organized according to principles beyond our evolutionary and cultural experience.

Beyond Anthropomorphism: The Mirror and the Other

THROUGHOUT HUMAN HISTORY, our approach to the non-human has oscillated between two problematic extremes: excessive anthropomorphism (projecting human qualities onto non-human entities) and excessive differentiation (denying meaningful commonality with non-human entities). Both extremes distort our understanding of non-human intelligence, whether terrestrial or potentially extraterrestrial.

Anthropomorphism attributes human motivations, emotions, and thought patterns to non-human entities, interpreting their behavior through human psychological frameworks. This tendency emerges naturally from our evolved psychology—particularly theory of mind mechanisms that evolved to predict other humans' behavior but get applied beyond their original domain. It allows us to make quick predictions about animal behavior but often leads to fundamental misinterpretations of non-human cognition.

Excessive differentiation, conversely, assumes radical incommensurability between human and non-human experience, treating non-human entities as wholly other. This approach, while avoiding anthropomorphic projection, often fails to recognize genuine commonalities emerging from shared evolutionary history, physical embodiment, or functional constraints. It can lead to denying consciousness or complex cognition in non-human entities despite evidence for these capacities.

A more balanced approach recognizes both commonality and difference—what philosopher Thomas Nagel called "the view from nowhere," attempting to conceptualize reality neither from a purely human perspective nor from an impossible "objective" standpoint but from an intermediate position that acknowledges both shared reality and divergent perspectives on that reality.

This balanced approach becomes particularly important for extraterrestrialist thinking. Anthropomorphic projection would assume that aliens would share human desires, fears, values, and cognitive patterns—an assumption almost certainly false for truly alien intelligence. Yet assuming complete otherness would make any recognition or communication theoretically impossible, rendering the search for extraterrestrial intelligence futile by definition.

The challenge becomes finding what philosopher Hans-Georg Gadamer called a "fusion of horizons"—a partial overlapping of conceptual frameworks that allows for some mutual comprehension without assuming complete transparency. This fusion requires recognizing both potential commonalities (physical embodiment, existence within the same universe and physical laws) and potential differences (evolutionary history, sensory systems, cognitive architecture).

Several approaches might facilitate this balanced perspective:

1. **Comparative cognition**: Studying diverse forms of intelligence on Earth to better understand which aspects of cognition might be convergent across different evolutionary pathways and which might be unique to specific lineages
2. **Multiple working hypotheses**: Maintaining several different interpretative frameworks simultaneously when analyzing potential evidence of extraterrestrial intelligence, rather than settling prematurely on single explanations
3. **Interdisciplinary interpretation**: Bringing diverse disciplinary perspectives—physics, biology, cognitive science, anthropology, philosophy, linguistics—to bear on questions of alien recognition, acknowledging that no single framework captures all relevant dimensions
4. **Iterative refinement**: Treating understanding as an ongoing process rather than a achieved state, with provisional interpretations subject to continuous revision as new information emerges

These approaches don't eliminate the fundamental tension between anthropomorphism and radical otherness, but they allow for navigating this tension more consciously. They acknowledge that understanding alien intelligence would be neither impossible nor straightforward but would require sustained engagement across the boundaries of different forms of embodied cognition.

Perhaps most importantly, a balanced approach would recognize that the search for extraterrestrial intelligence is inevitably also a mirror reflecting human self-understanding. Our conceptualization of potential alien cognition reveals our assumptions about what constitutes intelligence, consciousness, and meaningful existence. By becoming more aware of these assumptions, we might develop not just better frameworks for recognizing non-human intelligence but deeper understanding of human cognition in its specificity rather than its assumed universality.

This self-reflective dimension doesn't make the search for extraterrestrial intelligence merely an exercise in human psychology. The cosmos likely contains forms of organization beyond current human conception, potentially including entities we might meaningfully recognize as intelligent. But it does suggest that the search represents a complex interaction between human conceptual frameworks and cosmic reality rather than a straightforward detection problem.

In this interaction lies both the fundamental challenge and the profound potential of extraterrestrialist thinking. By attempting to conceptualize intelligence beyond human parameters, we push against the boundaries of our cognitive architecture while remaining inescapably within them—a paradoxical position that mirrors the fundamental human condition of being simultaneously part of the natural world and capable of reflecting on that world from a perspective that seems to transcend it.

Conclusion: The Boundaries of the Knowable

THIS EXPLORATION OF the epistemological limits of alien recognition returns us to a central theme of this book: the tension between the likely existence of complex organization beyond Earth and the profound constraints on human ability to recognize or comprehend such organization if it differs significantly from terrestrial patterns.

These limits don't render the search for extraterrestrial intelligence futile. They do, however, suggest the need for greater epistemological sophistication in how we conduct this search—moving beyond the implicit assumption that intelligence is a universal category that would be mutually recognizable across species boundaries to a more nuanced understanding of how thoroughly perception and cognition are shaped by specific evolutionary history and embodied experience.

This more sophisticated approach would acknowledge several key principles:

1. **Our detection methods inevitably filter for intelligence somewhat similar to our own**, potentially missing vast ranges of complex organization beyond human parameters
2. **Communication, if it occurred, would face barriers far more fundamental than mere translation**, potentially requiring the collaborative construction of new conceptual frameworks rather than the exchange of existing ones
3. **Partial understanding might be possible even where complete comprehension is not**, allowing for meaningful

engagement without assuming transparent access to alien cognition

4. **The search itself generates valuable knowledge regardless of its success in detecting aliens**, deepening human understanding of our cognitive limitations and conceptual frameworks

These principles suggest a approach to SETI and extraterrestrialist thinking characterized by intellectual humility, conceptual creativity, and methodological pluralism. Rather than assuming we know what we're looking for, such an approach would remain radically open to patterns beyond current human conception while developing increasingly sophisticated tools for distinguishing meaningful complexity from random processes.

The recognition limits explored in this chapter don't apply only to hypothetical aliens. They also shape human understanding of non-human intelligence on Earth—from cetaceans to cephalopods, from corvids to artificial intelligence. Developing better frameworks for recognizing and engaging with these more accessible forms of non-human cognition might provide conceptual resources for the greater challenge of potential extraterrestrial intelligence.

More fundamentally, the question of alien recognition invites reflection on the nature of human knowledge itself—its inescapable embeddedness in specific evolutionary history, biological embodiment, and cultural frameworks. This reflection need not lead to radical skepticism or relativism but rather to what philosopher Donna Haraway calls "situated knowledge"—understanding that emerges from specific perspectives while remaining accountable to reality beyond those perspectives.

Such situated knowledge acknowledges both the limits of human cognition and its genuine capacity to reach beyond itself—not to achieve a "view from nowhere" but to develop increasingly sophisticated awareness of how knowledge is shaped by its conditions of production. This awareness doesn't eliminate cognitive limitations but makes them explicit, potentially allowing for more sophisticated navigation of the boundaries between the knowable and the unknowable.

In the end, the search for extraterrestrial intelligence represents not just a scientific project but a philosophical one—an attempt to approach the boundaries of human understanding while recognizing that we cannot simply step beyond them. This paradoxical position, simultaneously acknowledging cognitive constraints and reaching past them, may be the most honest and potentially productive approach to the profound question of whether we share the cosmos with other minds—and whether we could recognize them if we do.

Chapter 8: Cultural Analysis of Alien Representation

"We have met the alien, and he is us." — Adapted from Walt Kelly

The Mirror of Otherness

In 1995, filmmaker Jodi Dean published *Aliens in America: Conspiracy Cultures from Outerspace to Cyberspace*, arguing that alien abduction narratives functioned as "a technology for articulating themes of difference, unity, truth, and deception within American political consciousness." Dean's analysis represented a significant turn in how scholars approached extraterrestrial narratives—not as claims to be verified or debunked, but as cultural texts revealing deeper social and psychological patterns within the societies that produced them.

This chapter continues this cultural analysis approach, examining how representations of aliens across media, discourse, and popular imagination function as complex projections of human concerns rather than genuine attempts to imagine truly alien beings. Where previous chapters have explored scientific and philosophical constraints on our conception of extraterrestrial life, this chapter examines the cultural constraints—the ways in which our visions of aliens inevitably reflect and refract human preoccupations, anxieties, and desires.

This analysis doesn't suggest that alien representations have no relationship to potential extraterrestrial realities. Science fiction authors, scientists, and other creators often make genuine attempts to imagine truly alien beings and societies based on scientific principles and philosophical speculation. But even these earnest efforts remain shaped by the cultural contexts in which they emerge, revealing as much about the humans doing the imagining as about what might actually exist beyond Earth.

By examining alien representations as cultural texts, we can gain insight into both the limitations of our extraterrestrialist imagination and the social functions these representations serve. Aliens in popular culture and discourse aren't merely entertainment or speculation—they perform important cultural work, providing symbolic resources for exploring human identity, processing social anxieties, imagining social alternatives, and negotiating boundaries between self and other. Understanding this cultural work offers another perspective on the anthropocentrism that pervades extraterrestrialist thinking and that may limit our ability to recognize truly alien intelligence.

Aliens as Cultural Projections

THE ALIENS THAT POPULATE science fiction, UFO accounts, and popular discourse rarely demonstrate genuine alienness. Despite sometimes exotic appearances, they typically display recognizably human motivations, emotions, social structures, and values. Even when authors or filmmakers attempt to create truly alien entities—like the crystalline structures in Stanisław Lem's *Solaris* or the heptapods in Ted Chiang's "Story of Your Life" (adapted as the film *Arrival*)—these beings still function within narratives structured around human concerns and comprehensible through human conceptual frameworks.

This human-centrism isn't an artistic failure but reflects deeper patterns in how humans use representations of the "other" to explore aspects of ourselves. Throughout human history, imagined beings beyond human experience—from gods and spirits to monsters and aliens—have served as projective surfaces for exploring human potentials, fears, and social dynamics. Aliens represent the latest iteration of this pattern, functioning as what psychologist Carl Jung called "projections of the collective unconscious"—symbolic expressions of cultural archetypes and societal preoccupations.

Different alien types clearly embody different human concerns:

1. **The benevolent, enlightened aliens** (as in *Close Encounters of the Third Kind* or *Star Trek*) project human aspirations toward peace, wisdom, and technological advancement—embodying what we might become at our best
2. **The monstrous, predatory aliens** (as in *Alien* or *Independence Day*) manifest fears of the destructive potential within human nature and anxiety about vulnerability to forces beyond our control
3. **The inscrutable, truly alien aliens** (as in *Solaris* or *2001: A Space Odyssey*) represent the limits of human understanding and the potential meaninglessness of the cosmos from a human perspective
4. **The humanoid aliens with minor differences** (as in *Star Trek* races like Vulcans or Klingons) function as thought experiments exploring specific aspects of human nature by isolating and amplifying particular traits

These types form a kind of grammar of extraterrestrial representation, with different alien figures serving different narrative and psychological functions. Their diversity suggests not different conceptions of actual extraterrestrial life but different aspects of human experience being explored through the alien as symbolic device.

The temporal patterns in alien representation reveal even more clearly their function as cultural projections. Alien narratives evolve to reflect changing social anxieties and preoccupations:

- In the 1950s, alien invasion narratives like *The Day the Earth Stood Still* and *Invasion of the Body Snatchers* reflected Cold War anxieties about communism, conformity, and nuclear annihilation.
- The 1960s and 70s brought more benevolent aliens in works like *2001: A Space Odyssey* and *Close Encounters of the Third Kind*, reflecting countercultural hopes for transcendence and cosmic consciousness.
- The economic uncertainties of the 1980s saw a turn toward aliens as corporate or capitalist metaphors in films like *They Live* and *Aliens*, which depicted extraterrestrials as exploitative forces destroying human community.
- The 1990s and early 2000s brought alien abduction narratives into mainstream discourse, reflecting anxieties about loss of agency and bodily autonomy in an increasingly complex technological society.
- More recent representations often focus on communication challenges (*Arrival*), questions of consciousness (*Under the Skin*), or environmental themes (*Avatar*), reflecting contemporary preoccupations with cross-cultural understanding, artificial intelligence, and ecological crisis.

These shifting representations track social concerns more closely than developments in scientific understanding of potential extraterrestrial life. They demonstrate how thoroughly alien narratives function as cultural technologies for processing collective anxieties and aspirations rather than as genuine attempts to imagine extraterrestrial realities.

Psychoanalytic Reading of Alien Abduction Narratives

AMONG EXTRATERRESTRIAL narratives, alien abduction accounts offer particularly rich material for cultural analysis. Beginning with Betty and Barney Hill's famous 1961 abduction claim and proliferating through the 1980s and 90s, these narratives describe human captivity by aliens who typically perform medical examinations or reproductive procedures before returning the abductees to Earth, often with missing time and fragmentary memories recovered through hypnosis.

While these accounts have been extensively analyzed for their potential evidential value regarding actual extraterrestrial contact, their consistency as cultural narratives—regardless of their factual status—invites deeper psychological and social analysis.

Psychoanalytic approaches to abduction narratives identify several consistent symbolic patterns:

1. **Boundary violation themes**: Abduction accounts consistently emphasize penetration of personal and bodily boundaries—aliens enter homes, paralyze victims, insert implants, and conduct reproductive procedures. These boundary violations mirror anxieties about personal agency and bodily integrity in contemporary society.
2. **Medical-reproductive focus**: Abduction narratives typically center on examination tables, invasive procedures, and reproductive scenarios including hybrid breeding

programs. These elements suggest processing of anxieties about modern medical technology, reproductive control, and genetic engineering.

3. **Temporal discontinuity**: "Missing time" and recovered memories represent psychological themes of dissociation and trauma—experiences existing outside normal temporal processing that can only be accessed through special techniques.
4. **Transformative aftermath**: Many abductees report life changes following their experiences, including environmental awareness, spiritual awakening, or sense of cosmic purpose. These elements suggest incorporation of the experience into meaningful personal narrative.

These patterns bear striking resemblance to other trauma narratives, particularly those involving medical procedures, sexual assault, or childbirth trauma. The psychoanalyst John Mack, who worked extensively with abductees, noted these parallels while also emphasizing the transformative spiritual dimensions many abductees attached to their experiences.

From a Jungian perspective, abduction narratives represent contemporary manifestations of much older mythic patterns involving contact with supernatural beings—from fairy abductions in European folklore to shamanic initiations involving dismemberment and reassembly by spirits. These parallels suggest that alien abduction narratives perform similar psychological functions to these older myths, providing symbolic frameworks for processing traumatic experiences and integrating them into meaningful personal and cultural narratives.

The feminist psychoanalyst Susan Clancy has argued that abduction narratives often provide frameworks for processing actual traumatic experiences that victims may not be able to directly confront. The alien context provides psychological distance that allows engagement with traumatic themes while protecting the psyche from direct confrontation with more immediate sources of trauma.

These psychoanalytic readings don't necessarily negate the lived reality of abduction experiences for those who report them. As anthropologist Paul Devereux notes, such experiences may represent genuine altered states of consciousness that feel entirely real to experiencers. But they do suggest that the content and structure of these experiences—the specifically alien nature of the entities encountered—draws from cultural symbolism that helps the mind organize and make meaning from these experiences.

Whether representing actual encounters, sleep paralysis, recovered memories of medical trauma, or other experiences, alien abduction narratives reveal how thoroughly cultural symbols shape even our most intimate and anomalous experiences. They demonstrate the power of the alien as a contemporary archetype through which humans process experiences of vulnerability, violation, and potential transformation.

Colonial Discourse Analysis of Invasion Scenarios

ALIEN INVASION NARRATIVES—FROM H.G. Wells' 1898 novel *The War of the Worlds* to contemporary blockbusters like *Independence Day*—represent one of the most persistent themes in extraterrestrial fiction. These narratives typically depict technologically superior aliens arriving on Earth with conquest, resource extraction, or extermination as their goals, with humans cast as desperate defenders of their homeland.

Colonial discourse analysis—an approach developed by scholars like Edward Said, Homi Bhabha, and Gayatri Spivak to examine how colonial relationships are represented in literature and culture—offers powerful tools for understanding these narratives. From this perspective, alien invasion stories often function as inversions or projections of colonial histories, allowing predominantly Western audiences to process colonial guilt, fear of retribution, or anxieties about their own societies' imperial behaviors.

H.G. Wells made this connection explicit in *The War of the Worlds*, directly comparing the Martian invasion of England to British colonial genocide in Tasmania:

> "Are we such apostles of mercy as to complain if the Martians warred in the same spirit? The Tasmanians, in spite of their human likeness, were entirely swept out of existence in a war of extermination waged by European immigrants... Are we such apostles of mercy as to complain if the Martians warred in the same spirit?"

This explicit parallel suggests that even the earliest major alien invasion narrative functioned partly as a way for British audiences to process their colonial history by experiencing a fictional version of colonization from the perspective of the colonized rather than the colonizer.

This pattern continues in more recent invasion narratives, which often position aliens as engaging in precisely the kinds of exploitation that characterized European colonialism:

1. **Resource extraction**: In films like *Independence Day* and *Battle: Los Angeles*, aliens come to Earth specifically to extract resources, mirroring colonial resource extraction

policies.

2. **Civilizational replacement**: Narratives like *The Body Snatchers* or *The Thing* feature aliens who replace humans while maintaining outward appearances, echoing anxieties about cultural assimilation and replacement.
3. **Technological asymmetry**: Invasion narratives emphasize vast technological gaps between humans and aliens, mirroring the technological advantages that enabled European colonial conquest.
4. **Justified resistance**: These narratives position violent resistance against technologically superior invaders as morally justified, potentially allowing audiences from colonial powers to identify with anti-colonial resistance.

The prevalence of these invasion narratives in American and European science fiction specifically suggests their function in processing colonial histories from the perspectives of societies that were primarily colonizers rather than colonized. They allow audiences to simultaneously experience the position of the colonized (through identifying with humanity under alien attack) while ultimately reinforcing human superiority and right to self-determination (through humanity's eventual triumph).

This doesn't mean alien invasion narratives are merely displaced colonial guilt. They serve multiple functions simultaneously, including exploring legitimate concerns about technological vulnerability and species extinction. But their persistent popularity in Western cultural contexts suggests they provide important psychological resources for processing unresolved cultural traumas and moral ambiguities around colonialism.

More recent invasion narratives sometimes attempt to engage colonial themes more explicitly, as in the film *District 9*, which directly parallels alien detention with South African apartheid, or *Avatar*, which reverses the typical dynamic by positioning humans as colonial invaders of an alien world. These more reflexive approaches suggest growing awareness of the colonial subtexts in alien narratives, though they often still resolve their stories through individualized moral choices rather than addressing systemic colonial dynamics.

Aliens as Embodiments of the "Other" in Western Thought

BEYOND SPECIFIC COLONIAL parallels, aliens in Western cultural representation function more broadly as embodiments of "the Other"—the philosophical concept developed by thinkers from G.W.F. Hegel to Emmanuel Levinas to Simone de Beauvoir to describe how identity forms through contrast with what it is not. Western philosophy has long recognized that the self defines itself partly through differentiation from others, and this dynamic operates at both individual and cultural levels.

In Western thought, various groups have historically been positioned as "Others" against which Western identity defined itself: the "Orient" in relation to Europe, Indigenous peoples in relation to settlers, women in relation to men, nature in relation to civilization. Aliens represent the ultimate extrapolation of this othering process—beings defined precisely by their non-humanness, their status as the absolute Other against which humanity as a whole can define itself.

This function appears clearly in how aliens are typically represented physically. Despite the vast possibilities for alien physiology suggested by evolutionary biology, extraterrestrials in Western visual media overwhelmingly possess characteristics that symbolically mark otherness within human cultural contexts:

1. **Exaggerated craniums** suggesting hyper-intellectualism divorced from embodiment and emotion, reflecting Western mind-body dualism
2. **Insect or reptilian features** drawing on deep-seated human aversions and historical associations of these features with dehumanization
3. **Amorphous or gelatinous forms** violating clear boundaries between self and environment, reflecting anxieties about identity dissolution
4. **Mechanical or technological hybridization** embodying fears about technology's threat to human uniqueness and autonomy

These representational patterns draw on established visual languages of otherness from Western cultural history rather than on scientific speculation about likely evolutionary paths on other worlds. They function by exaggerating features that have historically been used to mark human groups as "other" within Western contexts—racialized features, disability markers, or associations with animality.

The philosopher Jacques Derrida's concept of "the constitutive outside"—the excluded other that defines the boundaries of the included self—helps explain aliens' cultural function. Aliens embody what must be excluded from human identity for that identity to maintain coherence. Their otherness serves not primarily to imagine genuine alternatives to humanity but to reinforce human identity through contrast.

This othering function appears perhaps most clearly in the trope of alien-human hybrids, which feature prominently in both science fiction and abduction narratives. These hybrids typically create horror or unease precisely because they violate the human/alien boundary that structures these narratives. They represent what literary theorist Julia Kristeva calls "the abject"—that which disturbs identity, system, and order by not respecting borders, positions, or rules.

Yet alongside this othering function, alien narratives often perform a paradoxical secondary function: the domestication of difference. Many alien narratives, particularly in film and television, ultimately reveal the alien as comprehensible, sympathetic, or even as a reflection of humanity's better nature. From *E.T.* to *Arrival*, these narratives often end with recognition of commonality across difference, suggesting that the alien other can ultimately be incorporated into human understanding.

This tension between radical otherness and ultimate familiarity reveals a central contradiction in Western cultural approaches to difference. Aliens must be strange enough to function as effective others against which human identity can define itself, yet familiar enough that narratives about them remain comprehensible and emotionally resonant for human audiences. This contradiction may partly explain why truly alien aliens remain relatively rare in cultural representation—beings defined by genuine incommensurability with human experience would challenge the very frameworks through which we construct meaningful narratives.

Indigenous Perspectives on Extraterrestrial Narratives

WESTERN DISCOURSE ON extraterrestrial life often presents indigenous cultures as either particularly susceptible to "alien worship" (as in ancient astronaut theories) or as possessing special insight into cosmic visitors through traditional knowledge. Both approaches typically misrepresent indigenous perspectives while appropriating indigenous cultural elements for Western extraterrestrialist narratives.

In reality, indigenous perspectives on potential extraterrestrial beings vary widely across different cultures and often differ fundamentally from Western extraterrestrialist frameworks. These differences reveal how thoroughly Western extraterrestrialism emerges from specific cultural contexts rather than representing universal human responses to cosmic questions.

Many indigenous cosmologies already include non-human intelligences with whom humans maintain relationships. These beings—often translated into English as spirits, ancestors, or animal persons—exist within relational networks that include humans rather than representing radical others from beyond Earth. Unlike the sharp human/alien binary common in Western extraterrestrialism, many indigenous frameworks recognize a continuum of different types of persons, only some of whom are human.

For example, Lakota traditions speak of Star Relatives or Star Nations—beings connected to specific stars who maintain ongoing relationships with human communities. These beings aren't understood as biological entities from other planets in the Western scientific sense, but as different types of persons with whom humans share kinship and reciprocal responsibilities. Similar relational frameworks appear in many indigenous cosmologies worldwide, from Aboriginal Australian songlines connecting terrestrial and celestial domains to Amazonian perspectives on non-human forest entities.

These indigenous frameworks often approach non-human intelligence through relational rather than objective epistemologies—ways of knowing based on relationship and reciprocity rather than detached observation. Knowledge about non-human beings emerges through direct interaction, ceremonial practices, dreams, and other forms of engagement rather than through the distanced observation characteristic of Western scientific approaches to potential extraterrestrial intelligence.

The Anishinaabe scholar Heidi Kiiwetinepinesiik Stark describes how her tradition already conceptualizes multiple worlds and diverse intelligences without requiring the Western extraterrestrialist framework:

> "When Settler people ask us about aliens, they're startled when we just nod and say 'of course there are other beings.' But we're not confirming their UFO stories. We're saying our cosmology already acknowledges many kinds of persons beyond the human. We don't need spaceships to make sense of a universe filled with different intelligences."

These indigenous perspectives challenge several key assumptions of Western extraterrestrialism:

1. The sharp distinction between Earth and space as separate domains
2. The conceptualization of non-human intelligence as necessarily alien or other
3. The prioritization of technological detection over relational engagement
4. The assumption that contact with non-human intelligence would be a novel event rather than an ongoing reality

When indigenous communities do engage with Western UFO narratives or extraterrestrialist discourse, they often integrate these elements into existing cosmological frameworks rather than adopting Western interpretations wholesale. For instance, some Navajo (Diné) interpretations incorporate UFO experiences into traditional frameworks involving skinwalkers or other non-human entities, understanding these phenomena through cultural categories that predate Western extraterrestrialist discourse.

Unfortunately, Western extraterrestrial discourse frequently appropriates indigenous perspectives in distorted forms. Ancient astronaut theories, popularized by authors like Erich von Däniken and television programs like *Ancient Aliens*, routinely misrepresent indigenous archaeological sites, cultural practices, and oral traditions as "evidence" of prehistoric alien visitation. These misrepresentations typically assume that indigenous peoples could not have created complex architecture, astronomy, or cultural systems without extraterrestrial intervention—a fundamentally colonial perspective that denies indigenous knowledge and achievement.

This appropriation extends beyond ancient astronaut theories to broader patterns in Western extraterrestrialist thinking. Indigenous ceremonies have been reinterpreted as "protocols for alien contact." Traditional knowledge about celestial phenomena has been mined for supposed references to alien visitation. And indigenous critiques of Western environmental destruction have been repackaged as "warnings from space brothers" rather than being respected as legitimate knowledge traditions with their own validity.

These appropriations reveal how Western extraterrestrialism often functions as a form of cultural extraction—taking elements of indigenous knowledge while discarding their original contexts and significance. This pattern echoes broader colonial dynamics while specifically distorting indigenous perspectives on non-human intelligence to fit Western preconceptions.

Genuine engagement with indigenous perspectives on non-human intelligence would require moving beyond these appropriative practices toward recognition of indigenous intellectual sovereignty—the right of indigenous peoples to maintain, protect, and develop their own knowledge systems without reduction to Western frameworks. Such engagement might offer valuable resources for developing less anthropocentric approaches to potential extraterrestrial intelligence, but only if conducted through ethical relationships based on respect rather than extraction.

Gender, Race, and Otherness in Alien Representation

THE SYMBOLIC FUNCTIONS of aliens in cultural representation become particularly clear when examining how gender and race intersect with extraterrestrial imagery. These intersections reveal that alien representations often encode and occasionally challenge social hierarchies and identity categories within human society rather than genuinely imagining non-human alterity.

Gender plays a central role in many alien representations, with extraterrestrials often embodying exaggerated versions of gendered stereotypes or serving as vehicles for exploring gender beyond human binary systems:

1. **Hyperfeminized aliens** like the green-skinned women of *Star Trek* or the cat-like aliens of *Avatar* often represent exotic sexuality and reproductive potential, reflecting how femininity has historically been associated with nature, body, and otherness in Western thought.
2. **Hypermasculinized aliens** like the Predator or the Klingons embody exaggerated masculine

traits—aggression, honor codes, physical prowess—often representing both threat and fascinating alterity to the (presumed male) viewer.

3. **Non-binary or alternative gender system aliens**, less common but increasingly present in science fiction, function as thought experiments exploring gender beyond human binaries, as in Ursula K. Le Guin's *The Left Hand of Darkness* or Ann Leckie's *Ancillary Justice*.

These gendered representations rarely challenge fundamental Western concepts of gender itself. Even aliens with "third genders" or alternative reproductive systems typically remain legible within human gender frameworks rather than presenting genuine alternatives. Their function is often to explore gender variations within human conceptual systems rather than to imagine intelligence beyond gendered categories altogether.

Racial coding of aliens follows similar patterns, with extraterrestrials frequently embodying racial stereotypes or serving as metaphors for racial difference and conflict:

1. **Primitive/savage aliens** often incorporate visual and narrative elements drawn from colonial representations of Indigenous peoples, as with the Ewoks in *Star Wars* or the Na'vi in *Avatar*.
2. **Hyper-technological but emotional/spiritual aliens** frequently incorporate orientalist stereotypes about Asian cultures, as with the Vulcans in *Star Trek* or the Minbari in *Babylon 5*.
3. **Assimilation-threatening aliens** like the Borg or the Body Snatchers encode anxieties about immigration and cultural change that often parallel racial panic narratives.

These racial codings rarely appear as explicit one-to-one mappings but function through visual and narrative resonances that connect alien otherness to racialized otherness within human contexts. The alien body becomes a site for processing anxieties and fascinations around human racial difference while maintaining plausible deniability through its ostensibly non-human nature.

Science fiction authors from marginalized groups have increasingly engaged with these dynamics explicitly, using alien narratives to explore and challenge racial and gender hierarchies. Octavia Butler's *Xenogenesis* trilogy directly examines how perceptions of alien bodies reflect and refract human racial and gender categories. Samuel R. Delany's work consistently challenges assumptions about what bodies and desires might exist beyond Earth. N.K. Jemisin's *Broken Earth* trilogy uses alien entities to explore complex intersections of race, gender, and power.

These more reflexive approaches demonstrate how alien representation can potentially move beyond simply reinscribing human categories of difference. By deliberately engaging with how aliens have traditionally encoded human hierarchies, these works begin to imagine genuinely alternative forms of embodiment and consciousness that might exist beyond human categorization systems.

The critical race theorist Sylvia Wynter's concept of "genres of being human" offers a useful framework for understanding these dynamics. Wynter argues that what counts as "human" has been historically defined through exclusion, with racially and gender-marginalized groups positioned as less-than-fully-human. Alien representations often encode these "genres of humanity," with extraterrestrials embodying traits associated with marginalized human groups while maintaining the speculative distance that science fiction provides.

This encoding means that alien representations cannot be understood as purely speculative attempts to imagine non-human otherness. They function simultaneously as processing systems for human social differences and hierarchies—reflecting, reinforcing, and occasionally challenging how those differences are constructed and maintained. Any attempt to imagine truly alien alterity must contend with how thoroughly extraterrestrial representation has been shaped by these human social dynamics.

Case Studies from Science Fiction Across Cultures

WHILE WESTERN SCIENCE fiction dominates global media distribution, extraterrestrial narratives emerge from diverse cultural contexts worldwide, offering valuable comparative perspective on how different societies imagine potential cosmic others. These cross-cultural variations reveal the contingency of Western extraterrestrialist tropes while suggesting alternative approaches to conceptualizing potential non-human intelligence.

Soviet and post-Soviet science fiction has traditionally emphasized philosophical and ethical dimensions of alien contact over invasion scenarios or individual heroism. Works like Stanisław Lem's *Solaris* (Poland), the Strugatsky brothers' *Roadside Picnic* (USSR), and films like Andrei Tarkovsky's *Stalker* typically present aliens as fundamentally incomprehensible entities whose nature transcends human understanding. These narratives often focus on the limitations of human cognition rather than on conquest or communion, reflecting philosophical traditions emphasizing epistemological humility.

Chinese science fiction has experienced remarkable international recognition in recent years, particularly through Liu Cixin's

Three-Body Problem trilogy. This work presents a distinctive approach to extraterrestrial contact, emphasizing cosmic sociology—the study of civilizational interactions across vast spatial and temporal scales. Rather than focusing on individual encounters or invasion scenarios, Liu examines how civilizational logic might operate in a universe where resources are limited and detection equals vulnerability. This approach reflects distinctive Chinese historical experiences with civilizational continuity and long-term strategic thinking.

Japanese science fiction has developed unique approaches to human-alien relationships through media like anime and manga. Works such as *Neon Genesis Evangelion* blur boundaries between alien, human, and divine, while series like *Dragon Ball* integrate extraterrestrial narratives with spiritual and martial arts traditions. The filmmaker Hayao Miyazaki consistently creates non-human entities that defy simple categorization as aliens, spirits, or natural forces, reflecting Shinto and Buddhist influences that conceptualize consciousness as present throughout natural systems rather than limited to human-like entities.

African speculative fiction increasingly engages with extraterrestrial themes through distinctive cultural lenses. Nnedi Okorafor's *Lagoon* depicts first contact occurring in Lagos, Nigeria, deliberately centering African experiences and incorporating Yoruba cosmology into its alien conceptualization. Films like *District 9*, while directed by South African-born Neill Blomkamp, explicitly engage with South Africa's apartheid history through its alien narrative. These works challenge the predominantly Western settings and perspectives of mainstream extraterrestrial fiction.

Indigenous futurism, a movement encompassing Indigenous science fiction authors and artists worldwide, often reconceptualizes relationships with non-human entities beyond Western extraterrestrialist frameworks. Writers like Grace Dillon (Anishinaabe), Gerald Vizenor (Anishinaabe), and Darcie Little Badger (Lipan Apache) create narratives where the boundaries between human, alien, animal, and spirit are fluid and where relationship rather than difference defines interactions with non-human intelligence.

These diverse approaches don't simply offer "cultural flavor" to a universal extraterrestrialist framework but present fundamentally different conceptual systems for imagining non-human intelligence. They challenge core assumptions of Western extraterrestrialism:

1. The sharp binary between human and alien
2. The presumption that contact would be a singular, unprecedented event
3. The focus on conflict, invasion, or communion as primary contact narratives
4. The assumption that Western nations would naturally lead any contact scenario

Cross-cultural comparison reveals how thoroughly Western extraterrestrialism emerges from specific historical and cultural contexts—particularly colonial history, Judeo-Christian theological frameworks, and Cold War geopolitics. Alternative cultural traditions offer different resources for imagining cosmic relationships that don't necessarily center human exceptionalism or position the non-human as radical other.

These cross-cultural perspectives suggest that genuinely post-anthropocentric approaches to potential extraterrestrial intelligence might emerge not just from scientific or philosophical revisions to Western extraterrestrialism but from sustained engagement with diverse cultural frameworks worldwide. The limitations of Western extraterrestrialist imagination might be partially transcended not by rejecting cultural dimensions of extraterrestrial thinking but by expanding the cultural resources brought to bear on these questions.

Government Communication Strategies About Possible Alien Contact

MOVING FROM FICTIONAL to institutional representation, government communication about potential extraterrestrial contact reveals another dimension of how cultural frameworks shape extraterrestrialist discourse. While less immediately visible than popular culture representations, official protocols, scientific research programs, and public messaging about potential contact significantly influence both expert and public conceptions of extraterrestrial possibilities.

Various governments and international bodies have developed protocols for potential extraterrestrial detection or contact scenarios. These include:

1. The **SETI Post-Detection Protocol**, developed in 1989 by the International Academy of Astronautics and endorsed by the International Institute of Space Law, which outlines procedures for verifying and announcing potential SETI detections
2. **NASA's Office of Planetary Protection**, which develops guidelines for preventing biological contamination between Earth and other celestial bodies (though primarily focused on microbial rather than intelligent life)
3. The **UN Office for Outer Space Affairs**, which would theoretically coordinate international response to

confirmed detection, though with limited specific protocols established
4. Various **classified military protocols** for responding to unidentified aerial phenomena, recently partially revealed through US government UFO report declassifications

These institutional frameworks reveal significant cultural assumptions about how contact might occur and what forms it might take. They overwhelmingly focus on either radio signal detection or physical visitation scenarios, reflecting the dominant extraterrestrialist paradigm rather than more diverse possibilities for how non-human intelligence might manifest or communicate.

The rhetorical framing of these protocols typically emphasizes scientific validation, international cooperation, and information control—values reflecting the institutional cultures that produced them. They position potential contact primarily as a scientific and security event rather than as a cultural, philosophical, or spiritual development, revealing institutional priorities regarding which dimensions of potential contact are considered most significant.

Public-facing government communication about extraterrestrial possibilities has evolved significantly over time, reflecting changing institutional concerns and cultural contexts:

1. **Cold War era (1950s-1980s)**: Government communication typically dismissed UFO reports while maintaining classified research programs, reflecting security concerns about misidentified Soviet technology and public panic
2. **Post-Cold War period (1990s-2000s)**: NASA increasingly emphasized astrobiology and exoplanet research in public messaging, shifting focus from UFOs to

scientific search for microbial or primitive life

3. **Contemporary period (2010s-present)**: Multiple governments have released previously classified UFO/UAP information while maintaining careful agnosticism about origins, reflecting changing public discourse and institutional positioning

These evolving communication strategies reveal shifting cultural functions of extraterrestrialist discourse within government contexts. During the Cold War, extraterrestrial narratives potentially threatened national security by distracting from or confusing response to terrestrial threats. In the contemporary period, extraterrestrial possibilities serve more complex institutional functions—simultaneously justifying space research budgets, addressing public interest, maintaining scientific credibility, and managing security implications.

The sociologist Pierre Bourdieu's concept of "fields of cultural production" helps illuminate these dynamics. Extraterrestrialist discourse operates differently in different institutional fields—scientific, military, political, media—with each field applying its own values and frameworks to the possibility of extraterrestrial intelligence. These different fields compete to define the "legitimate" approach to extraterrestrial questions, with scientific institutions generally dominating official discourse while military and security perspectives often operate through classification and selective disclosure.

This institutional competition reveals that extraterrestrialist discourse serves not just cultural but political functions. Control over the "official" narrative regarding potential extraterrestrial contact represents significant symbolic power with material consequences for research funding, security policy, and international relations. The current careful positioning of most governments—acknowledging unexplained phenomena while avoiding claims about extraterrestrial origins—reflects this complex political dimension of extraterrestrialist discourse.

Government communication strategies also reveal significant cultural biases in how potential contact is imagined institutionally. Protocols overwhelmingly assume that Western scientific institutions and major world powers would lead any contact scenario, marginalizing potential roles for indigenous communities, religious institutions, or non-Western knowledge systems despite their potential relevance to interpreting genuinely novel phenomena. This institutional centralization reflects and reinforces the anthropocentrism and Western-centrism of conventional extraterrestrialist thinking.

More reflexive institutional approaches have occasionally emerged, particularly through scientific workshops exploring societal implications of potential contact. NASA's Astrobiology Institute and the SETI Institute have sponsored interdisciplinary dialogues incorporating anthropological, philosophical, and religious perspectives alongside natural sciences. These initiatives suggest growing recognition that potential contact would have dimensions beyond scientific validation or security response, though they remain relatively marginalized within institutional frameworks.

The Alien as Psychological Necessity

THE CULTURAL FUNCTIONS of alien representation explored throughout this chapter suggest that extraterrestrialist discourse serves essential psychological and social purposes beyond speculation about actual extraterrestrial possibilities. The alien, as cultural figure, performs work that human societies appear to need—providing symbolic resources for processing complex psychological and social dynamics that might otherwise remain unarticulated or unaddressed.

The psychoanalyst Jacques Lacan's concept of the "big Other"—the symbolic order through which human subjects understand themselves and their world—offers insight into this psychological function. The alien serves as a particular manifestation of otherness that allows human subjects and societies to constitute their identity through distinction from what they are not. The radical otherness of the extraterrestrial provides a conceptual horizon against which human commonality can emerge—a shared difference from the truly alien that potentially transcends differences within humanity.

This psychological function helps explain why alien representations persist even in the absence of empirical confirmation of extraterrestrial intelligence. The alien serves necessary symbolic functions regardless of whether actual extraterrestrials exist. It provides:

1. **Cognitive estrangement** - A perspective outside human experience that allows critical reflection on aspects of human society otherwise taken for granted
2. **Containment of cultural anxieties** - A symbolic vessel for processing fears about technology, otherness, and social change
3. **Exploration of human potential** - A projection screen for imagining what humanity might become, for better or worse
4. **Processing of trauma** - A symbolic framework for articulating experiences that exceed normal categorical boundaries
5. **Spiritual displacement** - A secular substitute for religious

frameworks addressing human cosmic significance

These functions suggest that alien representations might persist even if conclusive evidence demonstrated that no other intelligent life exists in our cosmic vicinity. The psychological need for the alien as symbolic resource might generate new forms of extraterrestrialist discourse adapted to whatever empirical constraints emerged, continuing to perform its cultural work through modified narratives.

This psychological necessity doesn't mean that extraterrestrial intelligence doesn't or couldn't exist, nor does it invalidate scientific search for such intelligence. Actual extraterrestrials, if detected, would exist independently of human psychological needs or cultural representations. But it does suggest that our approaches to both searching for and conceptualizing potential extraterrestrial intelligence remain deeply influenced by these psychological and cultural functions—a influence that may significantly constrain our ability to recognize or comprehend truly alien intelligence if it exists.

The challenge for more sophisticated extraterrestrialist thinking is not to eliminate these psychological dimensions—which would be neither possible nor desirable—but to develop greater awareness of how they shape our approaches to potential extraterrestrial intelligence. By recognizing the cultural and psychological work that alien representations perform, we might develop more nuanced frameworks that acknowledge these functions while remaining open to possibilities beyond our current extraterrestrialist imagination.

Conclusion: Aliens as Mirrors of Humanity

THIS CULTURAL ANALYSIS of alien representation returns us to a central insight: our visions of extraterrestrial intelligence reveal more about humanity than about what might actually exist beyond

Earth. The aliens we imagine—whether in science fiction, UFO accounts, scientific speculation, or institutional protocols—function primarily as mirrors reflecting human concerns, anxieties, desires, and social dynamics rather than as genuine attempts to conceptualize truly alien alterity.

This mirroring function doesn't render extraterrestrialist thinking worthless. These cultural projections perform important psychological and social work, providing resources for processing complex human experiences that might otherwise remain unarticulated. And genuine attempts to imagine truly alien possibilities exist alongside these projective functions, particularly in the work of speculative fiction authors, philosophers, and scientists who deliberately challenge anthropocentric assumptions.

But the dominance of human projection in extraterrestrialist discourse does suggest the need for greater reflexivity—awareness of how thoroughly our conceptions of potential extraterrestrial intelligence are shaped by human psychological needs and cultural contexts rather than by objective consideration of cosmic possibilities. Without this reflexivity, we risk mistaking our projections for genuine openness to extraterrestrial alterity, potentially limiting our ability to recognize truly alien intelligence if we encounter it.

More sophisticated approaches would acknowledge the inevitable cultural dimensions of extraterrestrialist thinking while developing practices that might partially transcend their limitations:

1. **Cultural comparative analysis** examining how different societies imagine extraterrestrial possibilities, revealing the contingency of Western extraterrestrialist frameworks
2. **Interdisciplinary dialogue** bringing diverse perspectives—scientific, philosophical, anthropological,

artistic, indigenous—to bear on questions of potential non-human intelligence

3. **Phenomenological approaches** exploring the structures of human experience that shape our conception of otherness and investigating how these structures might be temporarily bracketed or modified
4. **Artistic experimentation** using creative practices to imagine alterity beyond conventional representational frameworks
5. **Critical reflexivity** consistently examining how human psychological and social needs shape our extraterrestrialist discourse

These approaches wouldn't eliminate the cultural dimensions of extraterrestrialist thinking but might create greater space for genuinely encountering otherness beyond current human conception. They suggest an extraterrestrialism that acknowledges its own cultural foundations while remaining open to what might exist beyond them—an approach that uses awareness of human psychological and cultural patterns not to dismiss extraterrestrial possibilities but to engage them with greater sophistication.

In the end, the cultural analysis of alien representation doesn't suggest abandoning the search for extraterrestrial intelligence but approaching it with greater awareness of its psychological and cultural dimensions. By recognizing how thoroughly aliens function as mirrors of humanity, we might develop greater capacity to perceive what might exist beyond the reflection—the genuinely alien intelligences that may share our cosmos but transcend our current capacity to imagine them.

Chapter 9: Beyond Extraterrestrialism: Alternative Frameworks

"The most alien planet is Earth." — Ursula K. Le Guin

Reimagining the Cosmic Question

Previous chapters have examined how our conception of potential extraterrestrial intelligence is constrained by anthropocentrism embedded in scientific methodologies, physical models, biological assumptions, probability calculations, technological projections, religious frameworks, epistemological limitations, and cultural representations. These constraints don't merely limit our imagination but potentially restrict our ability to recognize truly alien forms of existence that might share our cosmos.

If conventional extraterrestrialist thinking is so thoroughly constrained by human-centered conceptual frameworks, what alternatives might exist? How might we approach the question of potential non-human cosmic intelligence without simply projecting human patterns onto the universe? This chapter explores philosophical, methodological, and conceptual frameworks that attempt to move beyond the limitations of conventional extraterrestrialism toward more genuinely open approaches to cosmic possibility.

This exploration isn't merely theoretical. Alternative frameworks for conceptualizing potential non-human intelligence might significantly reshape both scientific search strategies and philosophical understanding. They might help develop detection methods calibrated to recognize patterns currently invisible to conventional SETI approaches. They might provide conceptual resources for comprehending truly alien forms of existence if we encounter them. And they might offer more sophisticated approaches to humanity's own positioning within cosmic contexts, whether or not we share the universe with other intelligences.

The alternatives explored here don't represent a comprehensive survey of all possible approaches nor a definitive solution to extraterrestrialism's limitations. Rather, they constitute initial explorations of how we might begin thinking beyond current conceptual constraints—intellectual experiments in post-anthropocentric cosmic imagination. By examining these alternatives, we might develop greater awareness of the contingency of current extraterrestrialist frameworks and greater openness to possibilities that exceed them.

Post-Anthropocentric Approaches to Cosmic Existence

ANTHROPOCENTRISM—THE positioning of human experience, values, and perspectives as central or normative—pervades extraterrestrialist thinking at multiple levels. From defining life in Earth-specific terms to measuring intelligence against human cognitive capabilities to projecting human technological trajectories onto hypothetical civilizations, conventional approaches to extraterrestrial possibility remain thoroughly human-centered despite their cosmic aspirations.

Post-anthropocentric philosophies attempt to develop conceptual frameworks that decenter human experience without claiming impossible objectivity or neutral observation. Rather than pretending to escape human perspective entirely—which would be neither possible nor desirable—these approaches acknowledge human situatedness while developing practices that might partially transcend its limitations.

Several philosophical movements offer resources for such post-anthropocentric approaches to cosmic questions:

Posthumanism, developed by theorists like Donna Haraway, Rosi Braidotti, and Cary Wolfe, challenges the exceptional status granted to humanity in Western philosophical traditions. Unlike transhumanism, which seeks to transcend current human limitations through technological enhancement, posthumanism questions the coherence of "the human" as a category separate from other beings. It examines how the boundary between human and non-human is constructed and maintained rather than representing a natural division.

Applied to extraterrestrialist thinking, posthumanism suggests moving beyond questions of whether alien intelligence would be "like us" toward more fundamental reconsideration of the categories through which we conceptualize cosmic existence. Rather than projecting human/alien binaries onto the cosmos, posthumanist approaches might recognize continuities and differences across multiple forms of embodied existence without privileging human forms as normative.

Multispecies ethnography, developed by anthropologists like Anna Tsing, Eduardo Kohn, and Marilyn Strathern, extends ethnographic methods to examine how humans exist within complex relationships with other species rather than separate from them. This approach studies how multiple species co-create shared worlds through their interactions, challenging human exceptionalism while developing methodologies for understanding non-human ways of being.

Applied to extraterrestrialist questions, multispecies approaches might examine how different forms of intelligence emerge through specific environmental relationships rather than existing as abstract capacities. This framework suggests that intelligence should be understood not as a single quality possessed to greater or lesser degrees but as diverse adaptive strategies emerging through particular evolutionary and ecological contexts.

Ecological thinking, particularly deep ecology developed by philosophers like Arne Naess and David Abram, emphasizes the embeddedness of all beings within natural systems rather than positioning humans as separate observers or controllers of nature. This approach challenges subject/object divisions that position humans as conscious agents observing an inanimate universe, instead recognizing consciousness and agency as distributed throughout complex systems.

Applied to cosmic questions, ecological frameworks suggest conceptualizing intelligence not as a property of individual beings but as an aspect of complex systems—a quality emerging through relationships rather than existing in isolation. This approach might recognize intelligence in atmospheric systems, geological processes, or stellar dynamics that conventional frameworks categorize as non-conscious physical phenomena.

These post-anthropocentric approaches share several key principles relevant to reimagining extraterrestrialist thinking:

1. They recognize that all knowledge emerges from particular embodied perspectives rather than from nowhere
2. They challenge binary distinctions between human/non-human, conscious/non-conscious, and agent/environment
3. They emphasize relationships and interactions rather than isolated entities or properties

4. They acknowledge the agency and significance of non-human beings and systems
5. They recognize consciousness as potentially distributed through systems rather than contained within individual minds

While these approaches don't eliminate human perspective—which would be neither possible nor desirable—they provide resources for conceptualizing cosmic existence beyond the conventional extraterrestrialist framework. They suggest possibilities for recognizing forms of agency, intelligence, and significance that don't conform to human patterns or priorities but might nonetheless share our cosmic context.

Object-Oriented Ontology and Alien Being

OBJECT-ORIENTED ONTOLOGY (OOO), developed by philosophers like Graham Harman, Timothy Morton, and Ian Bogost, offers particularly relevant resources for reimagining extraterrestrial possibilities. This philosophical movement attempts to develop frameworks for conceptualizing existence beyond human access or understanding—for recognizing that things exist in ways that exceed our ability to comprehend them fully.

Unlike traditional philosophical approaches that prioritize human access to reality through consciousness, language, or social construction, OOO argues that objects exist independently of human perception or conceptualization. More radically, it suggests that objects "withdraw" from all relations—that no entity, human or non-human, can fully access or exhaust the reality of another entity. Each object exists surplus to any possible interaction with or observation of it.

This philosophical approach has profound implications for extraterrestrialist thinking:

1. **Recognition of necessary opacity**: OOO suggests that truly alien beings would necessarily exceed human comprehension—not because of contingent limitations that might be overcome but because all beings exceed total comprehension by any other being.
2. **Flattened ontology**: Rather than positioning humans as uniquely conscious subjects observing non-conscious objects, OOO develops a "flat ontology" where all entities—human, animal, vegetable, mineral, conceptual—exist with equal ontological status, though with different properties and capacities.
3. **Indirect access**: OOO proposes that relationships between entities always involve translation and mediation rather than direct access. This suggests that potential communication with alien intelligence would necessarily involve creative translation across fundamentally different realities rather than direct information exchange.
4. **Withdrawal from totalization**: The object-oriented concept of "withdrawal" suggests that no framework—scientific, philosophical, or otherwise—could ever totally capture or explain alien existence. Aspects would always remain inaccessible to human understanding.

Timothy Morton's concept of "hyperobjects"—entities massively distributed across time and space, exceeding conventional human scales of perception and understanding—offers particular insight for extraterrestrialist thinking. Morton describes climate change and nuclear radiation as hyperobjects—phenomena that exist at scales humans can model mathematically but cannot directly perceive or fully comprehend.

Applied to extraterrestrial questions, the hyperobject concept suggests potential alien intelligences might operate at temporal, spatial, or conceptual scales that make them fundamentally imperceptible to human observation despite their reality. They might exist as entities too vast, too distributed, too slow, or too fast for human cognitive systems to recognize as coherent beings rather than as background conditions or dispersed phenomena.

Ian Bogost's concept of "alien phenomenology" extends these insights, asking how entities experience their worlds from their particular perspective. Bogost uses computational metaphors to imagine how different objects—from cameras to trees to radio signals—"perceive" their environments according to their specific capacities. This approach doesn't claim to access non-human experience directly but uses speculative methods to imagine how different material configurations might engage with their surroundings.

Applied to extraterrestrialist questions, alien phenomenology offers methods for speculating about how truly alien entities might experience reality without claiming direct access to that experience. It suggests developing multiple speculative models of alien perception and cognition based on different potential biological, physical, or other configurations, without privileging human-like models as normative.

These object-oriented approaches don't resolve the fundamental paradox of trying to conceptualize truly alien existence from an inescapably human perspective. But they do provide conceptual tools for acknowledging this paradox while developing more sophisticated approaches to potential extraterrestrial realities—approaches that recognize the necessary limitations of human understanding without abandoning the attempt to reach beyond those limitations where possible.

Speculative Realism and the Limits of Correlation

THE BROADER PHILOSOPHICAL movement of Speculative Realism, which emerged in the early 21st century through thinkers like Quentin Meillassoux, Ray Brassier, and Iain Hamilton Grant, offers additional resources for reimagining extraterrestrialist thinking. This diverse philosophical approach united around criticism of what Meillassoux termed "correlationism"—the assumption that we can only meaningfully discuss the correlation between thought and being, never either in isolation from the other.

Correlationism underlies many post-Kantian philosophical approaches, which argue (in different ways) that we cannot access reality independent of human conceptual frameworks. Whether through Kantian categories of understanding, phenomenological intentionality, linguistic structures, or social construction, these approaches suggest that reality-as-experienced is always already shaped by human cognitive structures rather than representing mind-independent reality.

Speculative Realism challenges this correlationist assumption, attempting to develop philosophical approaches that can meaningfully engage with reality beyond human access while acknowledging the inevitable human positioning of such attempts. Different speculative realist thinkers pursue this goal through different strategies:

Meillassoux uses mathematics to access what he calls "the great outdoors"—reality independent of human consciousness—arguing that mathematical formalization allows thought to access necessity beyond human correlation with reality.

Ray Brassier employs scientific realism to argue that scientific knowledge progressively approaches mind-independent reality, even if complete access remains impossible.

Iain Hamilton Grant develops a neo-Schellingian naturephilosophy that positions human thought as one manifestation of nature's self-unfolding rather than as separate from nature.

These diverse approaches share commitment to the possibility of thinking beyond purely human frames of reference without claiming impossible objectivity or neutral observation. They suggest that philosophy can meaningfully engage with reality beyond human correlation while acknowledging the human positioning of such engagement.

Applied to extraterrestrialist questions, these speculative realist approaches suggest several promising directions:

1. **Ancestral time**: Meillassoux's concept of "ancestrality" addresses how we can meaningfully think about reality before human existence. This approach might help conceptualize truly alien temporalities—beings operating at time scales vastly different from human experience.
2. **Scientific realism**: Brassier's scientific realism suggests that progressive scientific understanding might approach aspects of alien reality even when that reality transcends direct human experience or conceptualization.
3. **Powers ontology**: Grant's emphasis on powers and potentials rather than actualized entities suggests focusing on what cosmic systems might do rather than what they currently appear to be—an approach potentially more

open to emergent and non-obvious forms of organization.

4. **Non-anthropocentric necessity**: Speculative realist approaches generally seek necessary structures in reality that exist independent of human conceptualization, potentially providing frameworks for recognizing patterns that transcend human cognitive biases.

These approaches don't eliminate the paradox of attempting to think beyond human thinking, but they do provide more sophisticated frameworks for acknowledging this paradox while pursuing cosmic questions. They suggest developing what philosopher François Laruelle calls "non-standard" methods—approaches that recognize their own limitations while attempting to think according to the real rather than imposing human categories onto reality.

For extraterrestrialist thinking, these frameworks suggest moving beyond both naive realism (assuming direct access to mind-independent reality) and pure constructivism (reducing reality to human conceptual frameworks). Instead, they propose developing methods that acknowledge the inevitable human positioning of cosmic inquiry while systematically attempting to minimize anthropocentric bias and remain open to realities that exceed human categories.

Multi-Species Ethnography as a Methodological Model

MOVING FROM PHILOSOPHICAL frameworks to concrete methodologies, multi-species ethnography offers particularly promising models for approaching potential extraterrestrial intelligence. This ethnographic approach, developed primarily through anthropological research, examines how multiple species co-create shared worlds through their interactions. It studies how different forms of being engage with environments and each other without assuming human perspectives as normative.

Traditional ethnography studies human cultural systems through immersive participant observation, seeking to understand social worlds from within while maintaining analytical perspective. Multi-species ethnography extends this approach to examine how humans exist within complex relationships with other species—from domestic animals to microbes to forest ecosystems—challenging human exceptionalism while developing methodologies for understanding non-human ways of being.

This approach has several key principles relevant to extraterrestrialist thinking:

1. **Attention to communication beyond language**: Multi-species ethnographers study how different species communicate through diverse modalities—chemical, gestural, auditory, tactile—without privileging human linguistic communication as the only significant form.
2. **Recognition of multiple forms of agency**: Rather than positioning humans as the only meaningful agents within environments, this approach recognizes how multiple species exercise different forms of agency in creating shared worlds.
3. **Focus on relationship rather than essence**: Multi-species approaches examine how beings emerge through relationships rather than possessing fixed, isolated essences—intelligence appears as a relational property rather than an internal quality.
4. **Methodological creativity**: Studying non-human beings requires developing creative research methods beyond conventional interviews or surveys, including innovative

observational techniques, technological mediations, and speculative interpretation.

The anthropologist Eduardo Kohn's work with the Runa people of Ecuador exemplifies this approach. In "How Forests Think," Kohn examines how the Runa engage with forest ecosystems not as passive resources but as complex communicative networks including plants, animals, spirits, and humans. He analyzes how different beings within these networks interpret signs according to their particular embodiment and evolutionary history, creating what he calls "trans-species pidgins"—partial communicative systems that allow for limited but real interaction across species boundaries.

Applied to extraterrestrialist questions, Kohn's approach suggests examining how potential alien intelligence might engage with environments and communications according to its particular embodiment and evolutionary history rather than assuming human-like perception or interpretation. It suggests developing what might be called "alien ethnography"—speculative approaches to how differently embodied beings might experience and interpret their worlds without assuming human-like subjectivity.

The anthropologist Anna Tsing's research on matsutake mushrooms offers another relevant model. In "The Mushroom at the End of the World," Tsing examines how these mushrooms participate in complex ecological, economic, and cultural systems spanning multiple species and geographical regions. Rather than studying mushrooms as passive objects of human activity, she analyzes how their specific biological properties and behaviors shape human and non-human worlds alike.

Applied to extraterrestrialist questions, Tsing's approach suggests examining how potential alien beings might participate in complex systems through their specific properties and behaviors rather than as isolated entities. It suggests recognizing how intelligence might emerge through systemic interactions rather than existing as an internal property of individual beings.

These multi-species ethnographic approaches don't eliminate the challenge of studying truly alien intelligence from an inevitably human perspective. But they do provide methodological models for moving beyond purely projective approaches—for systematically attempting to recognize and engage with alterity without reducing it to human terms. They suggest developing what philosopher Isabelle Stengers calls "ecologies of practices"—diverse methodological approaches that acknowledge their specific limitations while attempting to engage with realities beyond those limitations.

For SETI and extraterrestrialist thinking, these approaches suggest developing search methods calibrated to recognize patterns that don't resemble human intelligence but nonetheless represent significant organization beyond random processes. They suggest systematic creativity in imagining how intelligence might manifest in ways radically different from human cognition or technology, while acknowledging the inevitable human positioning of such imagination.

Indigenous Cosmologies and Their Potential Contributions

INDIGENOUS KNOWLEDGE systems worldwide offer sophisticated cosmological frameworks that often conceptualize non-human agency, intelligence, and personhood differently from Western scientific or philosophical traditions. These diverse cosmologies don't represent a singular alternative to Western extraterrestrialism but offer multiple frameworks for conceptualizing human relationships with non-human entities in ways that might inform more sophisticated approaches to potential extraterrestrial intelligence.

Indigenous cosmologies typically differ from conventional Western frameworks in several ways relevant to extraterrestrialist thinking:

1. **Recognition of multiple types of persons**, only some of whom are human, rather than sharp human/non-human binaries
2. **Relational rather than objective epistemologies**, acknowledging that knowledge emerges through relationship and engagement rather than detached observation
3. **Recognition of agency distributed throughout natural systems** rather than limited to entities with human-like consciousness
4. **Integration of material and spiritual dimensions** rather than their separation into distinct domains
5. **Place-based knowledge** emerging through specific environmental relationships rather than abstract universal principles

These characteristics don't represent exotic alternatives to "rational" Western frameworks but sophisticated knowledge systems adapted to specific environmental and cultural contexts. They offer valuable resources for developing more nuanced approaches to potential non-human intelligence beyond conventional extraterrestrialist paradigms.

The Anishinaabe scholar and astronomer Annette Lee describes how her tradition conceptualizes relationships with celestial bodies not as detached observation of inanimate objects but as ongoing reciprocal engagement with other-than-human persons:

> "In Anishinaabe star knowledge, the stars are our relatives. We don't just observe them—we maintain relationships with them through ceremony, story, and reciprocal responsibilities. This doesn't mean we don't understand their physical nature, but that physical understanding exists within a broader framework of relationship."

This approach doesn't position stars as literally human-like but recognizes them as entities with whom humans maintain significant relationships rather than as mere objects of scientific study. It suggests frameworks for engaging with potential extraterrestrial intelligence through relationship rather than mere detection or classification.

The Māori concept of *mauri*—the life force present in all things, whether traditionally classified as living or non-living—offers another relevant framework. As Māori scholar Mere Roberts explains:

> "*Mauri* isn't limited to what Western science categorizes as biological life. Mountains, rivers, stones, winds—all possess *mauri*. This isn't primitive animism but a sophisticated recognition that what Western science calls 'inanimate' matter participates in relationships and possesses significance beyond human utilization."

This concept suggests approaches to extraterrestrial possibilities that don't limit intelligence or significance to biological entities but recognize how various cosmic systems might possess qualities Western frameworks typically limit to living beings.

Indigenous approaches to communication with non-human entities also offer relevant models. Many traditions include protocols for engaging with other-than-human persons through various mediations—dreams, altered consciousness, ceremonial contexts, or specialized interpreters. These approaches acknowledge both the possibility of communication across profound difference and the necessary limitations of such communication, suggesting models for potential extraterrestrial contact that recognize partial communication as valuable without requiring complete mutual comprehension.

These indigenous frameworks don't represent pre-scientific superstition but sophisticated knowledge systems adapted to specific environmental and cultural contexts. They offer valuable resources for developing more nuanced approaches to potential non-human intelligence beyond conventional extraterrestrialist paradigms.

Importantly, engaging with these indigenous frameworks requires ethical relationship rather than extractive appropriation. Western science and philosophy have long histories of selectively incorporating indigenous knowledge while discarding its context and ignoring its practitioners. Genuine engagement requires recognizing indigenous intellectual sovereignty—the right of indigenous peoples to maintain, protect, and develop their knowledge systems on their own terms rather than as resources for Western consumption.

Such ethical engagement might involve collaborative research partnerships that respect indigenous protocols, citation and compensation practices that acknowledge indigenous intellectual contributions, and institutional support for indigenous knowledge-keepers to develop their own approaches to extraterrestrial questions. These practices wouldn't guarantee transcending the limitations of conventional extraterrestrialism, but they would expand the intellectual resources brought to bear on cosmic questions beyond those of dominant Western frameworks.

Dark Matter/Energy as Potential Matrices for Non-Standard Existence

MOVING FROM CONCEPTUAL frameworks to cosmic phenomena, dark matter and dark energy offer intriguing possibilities for reimagining potential extraterrestrial existence beyond conventional parameters. These phenomena—which together constitute approximately 95% of the universe's mass-energy content yet remain largely uncharacterized beyond their gravitational effects—represent aspects of cosmic reality that exceed current human understanding despite their apparent prevalence.

Dark matter appears to interact with normal matter primarily or exclusively through gravity, showing little or no electromagnetic, strong, or weak nuclear interaction. This limited interaction profile means that potential structures composed of or within dark matter would be largely invisible to our conventional observation methods, which rely primarily on electromagnetic radiation (light across various wavelengths).

Some theoretical physicists have speculated about the possibility of "dark matter life"—organized systems existing primarily within dark matter substrates rather than within conventional matter. The physicist Lisa Randall has hypothesized dark matter structures including disk-like concentrations within galaxies. While she doesn't specifically propose intelligence within these structures, their existence would provide potential substrates for organization beyond conventional matter.

The theoretical physicist Eugene Lim has more explicitly explored the possibility of dark matter life, examining whether dark matter particles could potentially form structures complex enough to support phenomena analogous to life. While current models suggest significant constraints on such possibilities, our limited understanding of dark matter leaves substantial room for speculation about organizational possibilities beyond current physics.

Dark energy presents even more profound possibilities and challenges for extraterrestrialist thinking. As the apparent force driving the accelerating expansion of the universe, dark energy represents approximately 68% of all cosmic mass-energy yet remains almost entirely mysterious in its fundamental nature. Current models variously characterize it as a cosmological constant, a scalar field, or a property of spacetime itself rather than a conventional form of matter or energy.

These mysterious cosmic components suggest several possibilities relevant to reimagining extraterrestrial existence:

1. **Substrate possibilities**: Dark matter might provide alternative substrates for organization, potentially supporting structures and processes radically different from those in conventional matter.
2. **Interaction limitations**: The limited interaction between dark and conventional matter might explain why potential intelligence within these domains would remain largely undetectable through conventional means.
3. **Organizational unknowns**: Our limited understanding of these phenomena leaves open vast possibilities for organizational principles beyond current scientific models.
4. **Scale differences**: Dark matter and energy operate at cosmic scales, suggesting potential forms of organization distributed across volumes far exceeding conventional biological parameters.

The physicist and science fiction author Greg Egan explores related possibilities in his novel "Schild's Ladder," depicting intelligence emerging within the very structure of spacetime rather than as entities composed of conventional matter. While speculative, such possibilities highlight how thoroughly conventional extraterrestrialist thinking remains constrained by assumptions about intelligence requiring structures similar to those found in Earth biology.

Contemporary physics offers additional speculative possibilities through concepts like quantum mechanical non-locality, vacuum energy fluctuations, or holographic principles. Phenomena like quantum entanglement suggest possibilities for information exchange that don't involve conventional spatial proximity or energy transmission, potentially enabling forms of organization beyond current detection capabilities.

These physical possibilities don't provide ready-made alternatives to conventional extraterrestrialist thinking. They remain highly speculative and constrained by current theoretical limitations. But they do suggest potential domains for intelligence beyond those typically considered in SETI research or extraterrestrial speculation—domains that might host forms of organization we currently lack conceptual or observational tools to recognize.

More importantly, these possibilities highlight how thoroughly conventional extraterrestrialist thinking remains constrained by focusing on intelligence similar to Earth life—entities composed of conventional matter, organized through biochemical processes, and interacting through electromagnetic radiation. By expanding consideration to include these more exotic physical possibilities, we might develop search strategies and conceptual frameworks better calibrated to detect truly alien forms of existence.

Developing New Vocabularies for Truly Alien

Phenomena

THE CONCEPTUAL AND physical frameworks explored above suggest possibilities for extraterrestrial existence beyond conventional parameters. But articulating these possibilities requires linguistic and conceptual resources that don't simply extend existing categories but develop new vocabularies for potentially unprecedented phenomena. This linguistic challenge represents perhaps the most fundamental constraint on reimagining extraterrestrialist possibilities.

The philosopher Ludwig Wittgenstein famously wrote, "The limits of my language mean the limits of my world." Our ability to conceptualize potential extraterrestrial realities remains constrained by the categories available within human languages—categories that evolved to describe terrestrial environments and human social realities rather than cosmic possibilities beyond human experience.

Current extraterrestrialist vocabulary relies heavily on terms like "intelligence," "life," "technology," and "civilization"—concepts deeply embedded in human history and experience. Even when qualified as "alien intelligence" or "non-carbon-based life," these terms carry anthropocentric assumptions that may constrain rather than expand our cosmic imagination. They position potential extraterrestrial realities as variations on human categories rather than as phenomena potentially requiring entirely new conceptual frameworks.

Developing more adequate vocabularies requires several complementary approaches:

1. **Linguistic innovation**: Creating new terms specifically designed to articulate possibilities beyond current categories, as speculative fiction authors like Ursula K. Le Guin and China Miéville have done in developing

languages and concepts for truly alien entities.

2. **Conceptual hybridization**: Combining concepts from diverse domains—quantum physics, complex systems theory, anthropology, indigenous knowledge systems—to create hybrid frameworks better suited to articulating phenomena that transcend conventional categories.
3. **Methodological pluralism**: Employing multiple descriptive systems simultaneously rather than seeking single definitive frameworks, acknowledging that truly alien phenomena might require multiple complementary descriptions rather than unified categorization.
4. **Negative theology**: Using apophatic approaches that define potential alien phenomena by what they are not rather than by potentially limiting positive descriptions, maintaining openness to possibilities beyond current conception.

The science fiction author Ted Chiang exemplifies these approaches in his novella "Story of Your Life" (adapted as the film "Arrival"). Chiang creates a fictional alien language operating according to non-linear temporal principles, allowing its users to experience time non-sequentially rather than as linear progression. This fictional construction demonstrates how different linguistic structures might articulate radically different ways of experiencing reality.

Beyond fiction, some practical approaches to developing more adequate vocabularies for potential extraterrestrial phenomena include:

1. **Multidisciplinary working groups** bringing together diverse experts—physicists, biologists, anthropologists, linguists, indigenous knowledge-keepers, science fiction authors—to develop more sophisticated conceptual

frameworks for extraterrestrial possibilities.

2. **Linguistic experiments** deliberately attempting to articulate possible phenomena beyond current categories, using techniques like neologism, metaphorical extension, or grammatical innovation to stretch conceptual frameworks.
3. **Database approaches** systematically cataloging diverse ways of conceptualizing intelligence, life, and organization across human cultures and knowledge systems to identify constraints and possibilities in current frameworks.
4. **Algorithmic assistance** using computational tools to help identify patterns in current extraterrestrialist discourse and suggest alternative frameworks less constrained by conventional categories.
5. **Intentional untranslatability** deliberately maintaining certain concepts in their original languages rather than translating them into dominant frameworks, preserving their specific contextual meanings and relationships.

These approaches won't eliminate the fundamental constraint of conceptualizing cosmic possibilities from an inescapably human perspective. But they might create greater conceptual space for recognizing possibilities beyond current frameworks—developing what philosopher Gaston Bachelard called "epistemological breaks" that allow thinking beyond established categories rather than merely extending them.

For practical extraterrestrialist research, these linguistic innovations suggest developing multiple parallel search strategies based on different conceptual frameworks rather than privileging approaches based solely on extensions of known Earth phenomena. They suggest systematically varying fundamental assumptions about what might constitute significant organization or intelligence beyond Earth rather than focusing exclusively on familiar markers like narrow-band radio signals or technosignatures resembling human technology.

Quantum Consciousness Theories and Cosmic Interconnection

QUANTUM MECHANICAL phenomena like entanglement, superposition, and non-locality have inspired various theories suggesting consciousness might operate through quantum rather than classical principles. While controversial within mainstream neuroscience, these theories propose that quantum effects in neural microtubules, coherent biophoton fields, or other biological structures might play essential roles in conscious experience.

Physicists like Roger Penrose and Stuart Hameroff have developed detailed models suggesting quantum coherence in neural microtubules creates conditions for consciousness beyond classical computation. The biophysicist Mae-Wan Ho has proposed that living organisms maintain quantum coherence through structured water and liquid crystalline properties. And various researchers have explored potential quantum aspects of phenomena like photosynthesis, suggesting quantum effects might be more significant in biological systems than previously recognized.

These theories remain speculative and contested. Many mainstream neuroscientists and physicists argue that the warm, wet environment of biological brains would cause quantum decoherence too rapidly for quantum effects to significantly influence neural processing. However, ongoing research continues to discover quantum effects in biological systems previously thought too "messy" for such phenomena, suggesting caution before dismissing quantum biological possibilities.

For extraterrestrialist thinking, these quantum consciousness theories suggest several intriguing possibilities:

1. **Alternative consciousness substrates**: If consciousness involves quantum processes rather than solely classical computation, it might emerge through substrates quite different from neural systems as we know them.
2. **Non-local consciousness**: Quantum non-locality might allow forms of consciousness distributed across space rather than localized in discrete bodies or brains.
3. **Temporal anomalies**: Quantum temporal phenomena like delayed-choice experiments suggest possibilities for consciousness operating according to non-linear temporal principles.
4. **Field consciousness**: Quantum field theories suggest possibilities for consciousness emerging through field properties rather than discrete particle interactions, potentially allowing vastly different scales of conscious organization.

These possibilities don't provide ready-made alternatives to conventional extraterrestrialist thinking but suggest directions for more speculative approaches to potential non-human intelligence. If consciousness operates partly through quantum principles, intelligent entities might be organized quite differently from the discrete, spatially localized beings assumed in most extraterrestrialist speculation.

Beyond specific theories about biological quantum consciousness, broader applications of quantum principles to cosmic questions suggest additional possibilities. Quantum entanglement demonstrates that particles once interacted remain connected regardless of spatial separation, suggesting a universe fundamentally characterized by relationship rather than isolation. Quantum field theories position particles as excitations in underlying fields rather than as discrete entities, suggesting reality may be more field-like than thing-like at fundamental levels.

The physicist David Bohm developed a philosophical interpretation of quantum mechanics emphasizing what he called the "implicate order"—an underlying wholeness from which apparently separate entities and phenomena emerge as temporary manifestations. This approach suggests consciousness might be an intrinsic aspect of reality at fundamental levels rather than an emergent property of specific complex systems.

Similarly, the concept of "panpsychism"—the philosophical position that consciousness or proto-consciousness exists as a fundamental property throughout reality rather than emerging only in certain complex systems—has experienced renewed interest among philosophers of mind like David Chalmers and Galen Strawson. This approach suggests consciousness might be ubiquitous rather than rare, manifesting in different forms and degrees throughout cosmic systems.

These quantum-informed approaches to consciousness and cosmic organization suggest extraterrestrialist possibilities beyond discrete, localized entities with brain-like information processing systems. They suggest intelligence might exist as aspects of field-like phenomena, as manifestations of underlying implicate orders, or as properties distributed throughout cosmic systems rather than contained within individual beings.

For practical SETI research, these possibilities suggest developing search methods sensitive to patterns of organization that don't resemble discrete technological civilizations but might represent significant non-random organization nonetheless. They suggest looking for field-like patterns, quantum coherence in seemingly random signals, or organizational principles operating across conventional categorization boundaries.

While highly speculative, these quantum-informed approaches highlight how thoroughly conventional extraterrestrialist thinking remains constrained by classical physics models emphasizing discrete entities, local causation, and linear temporality. By incorporating quantum principles more thoroughly into our cosmic imagination, we might develop frameworks better calibrated to recognize truly alien forms of existence that don't conform to classical expectations.

Conclusion: Toward Epistemological Humility in Cosmic Questions

THE ALTERNATIVE FRAMEWORKS explored in this chapter don't provide definitive answers to extraterrestrialist questions. They don't conclusively determine whether other intelligences exist in our cosmic vicinity or what forms they might take. But they do suggest that conventional extraterrestrialist thinking—with its focus on recognizably life-like entities using recognizably technology-like tools to create recognizably civilization-like structures—may represent only a narrow slice of cosmic possibilities.

These alternatives point toward what might be called epistemological humility in approaching cosmic questions—recognition of how thoroughly our conception of potential extraterrestrial intelligence is shaped by specifically human cognitive architecture, terrestrial evolutionary history, and Western cultural frameworks. This humility doesn't require abandoning the search for cosmic companions but does suggest approaching this search with greater awareness of its conceptual limitations and greater openness to possibilities beyond current frameworks.

Several principles emerge from these alternative approaches that might guide more sophisticated extraterrestrialist thinking:

1. **Multiplicity rather than universality**: Recognizing that no single framework—scientific, philosophical, or cultural—can adequately capture the full range of cosmic possibilities, suggesting the need for multiple complementary approaches rather than unified theory.
2. **Relationship rather than objectivity**: Acknowledging that knowledge emerges through specific relationships between knowers and known rather than through detached observation, suggesting approaches that recognize how human positioning shapes cosmic understanding.
3. **Process rather than entity**: Focusing on dynamic processes and relationships rather than discrete entities, suggesting attention to how intelligence might manifest through systemic properties rather than individual beings.
4. **Partial rather than complete understanding**: Accepting that human comprehension of truly alien phenomena would necessarily remain partial and provisional rather than complete or definitive, suggesting approaches that acknowledge inevitable limitations while remaining open to further development.
5. **Ethical rather than purely epistemological approaches**: Recognizing that how we conceptualize potential extraterrestrial intelligence reflects and influences how we relate to difference within human contexts, suggesting attention to the ethical dimensions of extraterrestrialist thinking.

These principles don't resolve the fundamental challenge of thinking beyond human categories from an inescapably human perspective.

But they do suggest more sophisticated approaches to this challenge—approaches that acknowledge the contingency of current frameworks while systematically attempting to expand beyond their limitations.

For practical SETI research, these principles suggest developing what might be called "post-anthropocentric" search strategies—methods calibrated to detect patterns of organization that don't necessarily resemble human technology or biology but nonetheless represent significant non-random phenomena. Such strategies might include:

1. **Agnostic pattern detection** focusing on identifying any non-random organization without presupposing specific forms this organization should take
2. **Multi-scale observation** examining phenomena at widely varying temporal and spatial scales rather than privileging human-like scales of organization
3. **Minority report approaches** deliberately developing and testing alternative frameworks rather than focusing exclusively on consensus models
4. **Indigenous collaboration** working with diverse knowledge systems to develop detection approaches beyond conventional Western scientific parameters
5. **Artistic-scientific partnerships** using creative practices to expand conceptual frameworks beyond current scientific categories

These approaches wouldn't guarantee detecting truly alien intelligence if it exists in forms radically different from human expectation. The challenge of recognizing intelligence unlike our own remains profound and potentially insurmountable in some cases. But they would expand the range of possibilities we might recognize beyond those assumed in conventional extraterrestrialist frameworks.

More fundamentally, these alternative approaches suggest that the most profound impact of genuinely open extraterrestrialist thinking might not be discovering alien others but developing more sophisticated understanding of human cosmic positioning—recognition of both the limits and possibilities of human knowledge regarding realities beyond terrestrial experience. This understanding represents not retreat from cosmic questions but more nuanced engagement with them—engagement that acknowledges human particularity while remaining open to possibilities beyond current human conception.

In the final analysis, moving beyond extraterrestrialism doesn't mean abandoning the search for potential cosmic companions but approaching this search with greater awareness of its conceptual constraints and greater openness to possibilities beyond current frameworks. It means developing what philosopher Donna Haraway calls "situated knowledge"—understanding that emerges from specific perspectives while remaining accountable to realities beyond those perspectives, neither claiming impossible objectivity nor retreating to pure relativism.

This epistemological position—simultaneously acknowledging the constraints of human perspective and attempting to reach beyond those constraints where possible—may represent the most sophisticated approach to cosmic questions currently available to us. It offers no guarantees of successful contact or comprehensive understanding but does provide frameworks for more genuinely open engagement with what may exist beyond current human conception—frameworks that might help us recognize truly alien companions if they share our cosmic neighborhood, even if they take forms quite different from those imagined in conventional extraterrestrialist thinking.

Chapter 10: Conclusion: How Can We Know Aliens Exist?

"The most incomprehensible thing about the universe is that it is comprehensible." — Albert Einstein

The Limits of Our Cosmic Vision

Throughout this book, we have examined the conceptual frameworks through which humans imagine, search for, and theorize about extraterrestrial intelligence. From scientific methodologies to philosophical assumptions, from religious doctrines to cultural projections, we have explored how thoroughly our conception of potential cosmic companions is shaped by human experience, terrestrial biology, and Earth-bound cultural frameworks.

This critical examination hasn't aimed to dismiss the possibility of extraterrestrial intelligence or to discourage the search for such intelligence. Rather, it has sought to recognize the conceptual constraints that limit our cosmic imagination and to develop more sophisticated approaches that might partially transcend these constraints. It has attempted to distinguish between genuine openness to cosmic otherness and subtle projection of human patterns onto the universe at large.

The central insight emerging from this examination is both simple and profound: our vision of potential extraterrestrial intelligence tells us more about ourselves than about what might actually exist beyond Earth. The aliens we imagine—whether in scientific speculation, religious doctrine, or cultural representation—function primarily as mirrors reflecting human concerns, anxieties, desires, and limitations rather than as genuine attempts to conceptualize truly alien alterity.

This mirroring function doesn't render extraterrestrialist thinking worthless. Human projection serves important psychological and social purposes, providing conceptual resources for processing complex experiences and imagining possibilities beyond current reality. And genuine attempts to transcend anthropocentrism exist alongside these projective functions, particularly in the work of scientists, philosophers, and artists who deliberately challenge human-centered assumptions.

But recognizing the dominance of anthropocentric projection in extraterrestrialist discourse does suggest the need for greater reflexivity—awareness of how thoroughly our conceptions of cosmic intelligence are shaped by terrestrial experience rather than by objective consideration of cosmic possibilities. Without this reflexivity, we risk mistaking our projections for genuine openness to extraterrestrial alterity, potentially limiting our ability to recognize truly alien intelligence if we encounter it.

This conclusion returns us to the question posed in this chapter's title: How can we know aliens exist? The careful reader will have recognized that this question contains ambiguities requiring clarification before meaningful response becomes possible. What precisely do we mean by "know"? What constitutes "aliens" within our conceptual framework? What would count as "existence" for entities potentially unlike anything in human experience?

These ambiguities aren't merely semantic quibbles but reflect fundamental uncertainties at the heart of extraterrestrialist thinking—uncertainties that cannot be resolved through more precise definition because they emerge from the inescapable limitations of human cognition when confronted with potential realities beyond terrestrial experience. They reflect what might be called the constitutional uncertainty of cosmic questions—the recognition that certain forms of knowledge may remain fundamentally inaccessible to beings evolved for life on Earth rather than for cosmic comprehension.

This constitutional uncertainty doesn't render the search for extraterrestrial intelligence futile but does suggest approaching this search with greater awareness of its conceptual limitations and greater openness to possibilities beyond current frameworks. It suggests developing what philosopher Donna Haraway calls "situated knowledge"—understanding that emerges from specific perspectives while remaining accountable to realities beyond those perspectives, neither claiming impossible objectivity nor retreating to pure relativism.

Synthesis: What Have We Learned?

THROUGHOUT THE PRECEDING chapters, we have developed a critique of conventional extraterrestrialist thinking across multiple dimensions. Each chapter has examined specific aspects of how human cognitive biases, disciplinary assumptions, and cultural frameworks constrain our cosmic imagination and potentially limit our ability to recognize genuinely alien forms of intelligence. Together, these critiques suggest a comprehensive reassessment of how we conceptualize potential extraterrestrial existence.

In Chapter 1, we examined the historical and conceptual foundations of modern extraterrestrialism, exploring how this paradigm emerged from specific cultural and scientific contexts rather than representing a universal approach to cosmic questions.

We recognized how modern SETI parameters, alien representations, and extraterrestrial hypotheses reflect cultural assumptions specific to post-Enlightenment Western societies rather than neutral frameworks for cosmic understanding.

Chapter 2 critiqued biological assumptions in astrobiology, arguing that our conception of "life" itself reflects Earth-specific evolutionary history rather than universal categories. We explored how carbon chauvinism, observer bias, and anthropocentric definitions of life might blind us to radically different forms of cosmic organization, from alternative biochemistries to non-metabolic forms of complex organization.

In Chapter 3, we challenged probabilistic arguments for alien abundance, examining how the Drake Equation and similar frameworks transform limited data into seemingly rigorous predictions through mathematical operations that mask their anthropocentric assumptions. We recognized how Fermi's Paradox functions less as an objective cosmic puzzle than as a reflection of human conceptual limitations when imagining potential civilizational trajectories.

Chapter 4 critiqued technological projections like the Kardashev Scale and Dyson Sphere concepts, revealing how these frameworks project industrial capitalist values onto cosmic scales rather than considering diverse developmental possibilities. We explored alternative civilizational trajectories that might prioritize efficiency, sustainability, or consciousness enhancement over physical expansion and resource exploitation.

In Chapter 5, we examined physical and biological constraints on interstellar travel, recognizing how the inherent hostility of space to evolved biological organisms might fundamentally limit direct contact between intelligent species regardless of technological development. We considered how these constraints might produce cosmic isolation even in a universe teeming with intelligence.

Chapter 6 evaluated religious frameworks for extraterrestrial existence, recognizing how theological approaches remain constrained by anthropocentric assumptions despite claims to transcendent knowledge. We explored how religious impulses persist in ostensibly secular scientific cosmologies, revealing the deeply human psychological needs that shape all cosmic narratives.

In Chapter 7, we confronted epistemological limits to alien recognition, examining how human cognitive architecture, terrestrial evolutionary history, and embodied experience might fundamentally constrain our ability to recognize intelligence unlike our own. We considered how language games, phenomenological constraints, and observer bias might create recognition barriers between radically different forms of intelligence.

Chapter 8 analyzed cultural representations of aliens, revealing how extraterrestrials in fiction, discourse, and popular imagination function primarily as psychological projections rather than genuine attempts to imagine cosmic otherness. We explored how aliens serve as screens for processing human anxieties about technology, otherness, and social change rather than as serious engagements with potential cosmic reality.

In Chapter 9, we explored alternative frameworks beyond conventional extraterrestrialism, from object-oriented ontology to indigenous cosmologies to quantum consciousness theories. We considered how these diverse approaches might help transcend the limitations of standard extraterrestrialist thinking, developing more sophisticated frameworks for conceptualizing cosmic possibilities beyond anthropocentric projection.

Across these diverse critiques, several consistent patterns emerge:

1. **Projection rather than perception** dominates extraterrestrialist thinking, with human patterns, values, and concerns projected onto cosmic scales rather than genuine openness to alterity
2. **Disciplinary constraints** limit cosmic imagination within specific fields, with methodological assumptions often mistaken for objective reality rather than recognized as contingent approaches
3. **Binary thinking** pervades extraterrestrialist discourse, with human/alien, living/non-living, and intelligence/non-intelligence treated as natural categories rather than constructed distinctions
4. **Scale limitations** restrict imagination to human-like temporal and spatial parameters rather than considering vastly different scales of organization and development
5. **Technological determinism** shapes civilizational models, assuming universal trajectories based on human industrial development rather than diverse adaptive possibilities

These patterns don't represent mere theoretical concerns but practical limitations that potentially restrict both scientific search methods and philosophical understanding. They suggest that conventional approaches to extraterrestrial intelligence might systematically miss genuinely alien forms of existence by searching primarily for patterns that resemble human civilization rather than remaining open to alternative possibilities.

However, these critiques don't lead to pure relativism or abandonment of the search for extraterrestrial intelligence. Rather, they suggest more sophisticated approaches characterized by greater awareness of human cognitive limitations, greater methodological diversity, and greater openness to possibilities beyond current conception. They suggest an extraterrestrialism reformed through critical self-awareness rather than abandoned as inherently flawed.

Methodological Reforms in Astrobiology and SETI

IF CONVENTIONAL EXTRATERRESTRIALIST approaches remain constrained by anthropocentrism, what alternatives might better serve the search for potential cosmic companions? How might scientific research programs be reformed to increase their sensitivity to truly alien forms of existence rather than focusing primarily on variations of terrestrial patterns? Several methodological reforms suggest themselves:

1. **Agnostic biosignature research** focusing on identifying any atmospheric or environmental patterns that indicate non-random organization rather than specifically Earth-like biological processes. Rather than searching primarily for oxygen-methane disequilibria or other Earth-analog signatures, this approach would develop methods for recognizing any persistent pattern requiring explanation beyond abiotic processes, regardless of its similarity to Earth biochemistry.
2. **Multiple working hypotheses** as standard practice in extraterrestrial research, deliberately developing and testing diverse frameworks rather than concentrating primarily on consensus models. This approach would systematically vary fundamental assumptions about what might constitute life, intelligence, or civilization, testing multiple detection strategies without privileging Earth-analog approaches.
3. **Anomaly-focused observation** that prioritizes unexplained phenomena rather than confirmed patterns,

systematically investigating cosmic anomalies without premature categorization as either natural processes or artificial signals. This approach recognizes that truly novel phenomena might initially appear as errors, artifacts, or inexplicable noise rather than as clearly identifiable patterns.

4. **Interdisciplinary interpretation teams** bringing together diverse expertise—astronomy, biology, anthropology, philosophy, indigenous knowledge, creative arts—to examine potential extraterrestrial signals or phenomena. This approach acknowledges that recognition of truly novel patterns might require multiple interpretive frameworks rather than specialized expertise in any single discipline.
5. **Algorithmic agnosticism** developing artificial intelligence systems trained to recognize patterns without human-imposed categories, potentially identifying organizations that human perception might miss due to cognitive biases. These systems wouldn't eliminate human interpretation but might expand the range of patterns initially recognized as potentially significant.

These methodological reforms wouldn't guarantee detecting truly alien intelligence if it exists in forms radically different from human expectation. The challenge of recognizing intelligence unlike our own remains profound and potentially insurmountable in some cases. But they would expand the range of possibilities we might recognize beyond the relatively narrow parameters assumed in conventional approaches.

More fundamentally, these reforms suggest shifting the central question of extraterrestrial research from "Are we alone?" to "What exists?" This shift decenters human concerns about cosmic uniqueness or companionship in favor of genuine curiosity about cosmic reality regardless of its relationship to human existence. It moves from an anthropocentric framing focused on human cosmic status to a more open investigation of cosmic possibility.

This shift doesn't eliminate human perspective—which would be neither possible nor desirable—but it does position that perspective more explicitly as one vantage point among potential others rather than as the unmarked norm against which cosmic reality is measured. It acknowledges what philosopher Thomas Nagel called "the view from nowhere" as an impossible ideal while still attempting to move beyond purely human-centered frameworks where possible.

For practical SETI research, this approach suggests expanding beyond conventional narrowband radio searches and optical SETI to include more diverse search strategies:

1. **Information theory approaches** examining any cosmic signal for complexity beyond random noise or simple natural processes, regardless of apparent intent or recognizable meaning
2. **Non-electromagnetic searches** considering gravitational waves, neutrinos, cosmic rays, or other potential communication channels beyond the electromagnetic spectrum
3. **Pattern analysis across scales** looking for organization at extreme temporal and spatial scales rather than privileging human-like rhythms and structures
4. **Distributed phenomenon investigation** considering the possibility of intelligence manifesting through systemic properties rather than discrete technological signatures
5. **Non-communication technosignatures** examining potential indicators of intelligence beyond deliberate

communication attempts, from waste heat to elemental distributions to isotopic anomalies

These expanded search strategies wouldn't replace conventional SETI methods, which remain valuable within their parameters. Rather, they would complement these methods with approaches calibrated to detect patterns that conventional searches might miss, creating a more comprehensive program less constrained by anthropocentric assumptions about how intelligence would manifest across cosmic scales.

Philosophical Approaches That Embrace Radical Uncertainty

BEYOND SPECIFIC METHODOLOGICAL reforms, genuinely sophisticated extraterrestrialist thinking requires philosophical frameworks that can embrace radical uncertainty without retreating to either false certainty or complete skepticism. It requires approaches that acknowledge the fundamental limitations of human cosmic understanding while remaining genuinely open to possibilities beyond current conception.

Several philosophical traditions offer resources for developing such approaches:

Phenomenology, particularly as developed by philosophers like Maurice Merleau-Ponty and Edmund Husserl, examines how conscious experience is structured through embodied perception rather than abstract reason. This tradition recognizes how thoroughly cognition emerges through specific bodily engagement with environments rather than representing disembodied rational processes. Applied to extraterrestrialist questions, phenomenology suggests recognizing how human perception and cognition inevitably shape cosmic understanding while developing practices that might partially transcend these constraints through deliberate variation of perspective.

Pragmatism, in the tradition of William James, John Dewey, and Richard Rorty, evaluates concepts and theories based on their practical consequences rather than their correspondence to absolute truth. This approach treats knowledge as tool-like rather than mirror-like, valuable for specific purposes rather than as perfect representation of independent reality. Applied to cosmic questions, pragmatism suggests developing multiple conceptual frameworks calibrated for different purposes rather than seeking single definitive models, recognizing how different approaches might reveal different aspects of cosmic reality.

Non-Western epistemologies, including Indigenous knowledge systems worldwide, often emphasize knowledge as emerging through relationship and practice rather than through detached observation. These diverse traditions typically recognize multiple ways of knowing—experiential, narrative, ceremonial, observational—without privileging any single approach as exclusively valid. Applied to extraterrestrialist thinking, these epistemologies suggest methodological pluralism embracing diverse knowledge practices rather than restricting valid approaches to conventional scientific methods alone.

Feminist epistemology, particularly standpoint theory developed by philosophers like Sandra Harding and Donna Haraway, examines how knowledge emerges from specific social positions rather than from nowhere. This tradition argues that all knowledge reflects particular standpoints while developing approaches for more objective understanding through explicit acknowledgment of positioning. Applied to cosmic questions, feminist epistemology suggests "strong objectivity" that acknowledges how human evolutionary history, terrestrial embodiment, and cultural frameworks shape cosmic understanding rather than claiming neutral observation.

These diverse philosophical approaches share several principles relevant to reformed extraterrestrialist thinking:

1. They recognize knowledge as positioned rather than universal, emerging through specific relationships between knowers and known rather than representing view-from-nowhere objectivity
2. They value multiple ways of knowing rather than privileging single methodologies, acknowledging how different approaches might reveal different aspects of complex realities
3. They emphasize practice and engagement rather than abstract theorization alone, recognizing knowledge as emerging through active relationship rather than passive observation
4. They acknowledge both the possibilities and limitations of human understanding, neither claiming comprehensive knowledge nor surrendering to complete skepticism

Together, these philosophical resources suggest what might be called "epistemological humility" in approaching cosmic questions—recognition of the inevitable limitations of human understanding coupled with commitment to reaching beyond those limitations where possible. This humility doesn't render extraterrestrialist inquiry futile but suggests approaching it with greater awareness of its inherent constraints and greater openness to possibilities beyond current frameworks.

For practical extraterrestrialist research, these philosophical approaches suggest specific practices:

1. **Explicit acknowledgment of assumptions** in research design and interpretation, recognizing how methodological choices shape what phenomena become visible or remain undetectable
2. **Deliberate variation of perspective** through interdisciplinary collaboration, cultural diversity, and methodological pluralism rather than consensus enforcement
3. **Negative results publication** treating non-detection as valuable scientific information rather than as failure, building cumulative understanding of search constraints
4. **Ongoing conceptual innovation** deliberately developing new frameworks rather than working exclusively within established parameters
5. **Collaborative interpretation** bringing diverse perspectives to bear on ambiguous data rather than privileging single disciplinary frameworks

These practices wouldn't eliminate the fundamental limitations of human cosmic understanding but might create greater space for recognizing possibilities beyond current extraterrestrialist imagination. They suggest an approach characterized by genuine curiosity about cosmic reality coupled with sophisticated awareness of how human positioning shapes that curiosity and its potential satisfaction.

The Ethical Dimensions of Alien Conceptualization

THE QUESTION OF POTENTIAL extraterrestrial intelligence contains not only scientific and philosophical dimensions but ethical ones as well. How we conceptualize possible cosmic others reflects and influences how we relate to difference within human contexts, making extraterrestrialist thinking ethically significant regardless of whether actual aliens exist or whether we ever contact them.

Several ethical dimensions deserve particular attention:

1. **Representation ethics**: How we represent potential aliens—whether as threats, resources, saviors, or companions—reflects broader patterns in how human societies conceptualize otherness. Extraterrestrialist discourse that positions aliens primarily as threats to be feared or resources to be exploited reinforces similar patterns in approaching human difference, while frameworks emphasizing potential for mutual learning and cooperation might support more ethical approaches to human diversity.
2. **Projection awareness**: When extraterrestrialist thinking projects terrestrial power dynamics—colonial relationships, resource exploitation, technological domination—onto cosmic scales, it potentially naturalizes these dynamics as inevitable rather than as contingent human patterns requiring ethical evaluation. Critical awareness of these projections allows more ethical consideration of power relationships in both cosmic and terrestrial contexts.
3. **Communication ethics**: Protocols for potential alien contact—whether through deliberate messaging or

response to detected signals—raise ethical questions about who speaks for humanity, what values are communicated, and what relationships are proposed. These protocols require ethical reflection on human representation, value pluralism, and relationship formation rather than merely technical consideration of effective transmission.

4. **Environmental ethics**: How we conceptualize potential alien environments—as resources for human use, as objects of scientific study, or as systems with their own integrity—reflects broader environmental ethics applicable to terrestrial contexts as well. Extraterrestrialist thinking that recognizes the potential value of alien environments beyond human utility might support similar recognition for Earth's ecosystems.
5. **Existential ethics**: Perhaps most fundamentally, how we approach potential cosmic intelligence reflects basic ethical orientations toward existence itself—whether the cosmos is conceptualized primarily as resource for human use, as object of human knowledge, or as reality with significance beyond human concerns. These orientations shape not only extraterrestrialist thinking but human relationships with all aspects of reality.

These ethical dimensions suggest that extraterrestrialist discourse serves as ethical practice regardless of whether actual aliens exist or are ever contacted. How we imagine, represent, and potentially communicate with cosmic others constitutes an ethical position toward otherness that transcends the specific question of extraterrestrial intelligence to encompass broader questions of relationship across difference.

For practical extraterrestrialist research, these ethical considerations suggest specific approaches:

1. **Representation diversity** in both scientific and cultural extraterrestrialist discourse, deliberately varying how potential aliens are conceptualized rather than defaulting to threat narratives or resource frameworks
2. **Inclusion of marginalized perspectives** in contact protocol development, ensuring that human messages or response strategies incorporate diverse cultural values rather than representing narrow Western frameworks alone
3. **Environmental consideration** in both observation strategies and potential contact scenarios, recognizing the potential value of alien environments beyond their utility for human purposes
4. **Critical self-awareness** about how extraterrestrialist frameworks reflect and potentially reinforce terrestrial power dynamics, with deliberate attention to avoiding colonial or exploitative patterns in cosmic conceptualization
5. **Value pluralism** in assessing potential cosmic intelligence, remaining open to forms of significance beyond human frameworks rather than measuring all cosmic phenomena against human values alone

These approaches wouldn't resolve all ethical questions raised by potential extraterrestrial intelligence but would constitute more ethically reflective engagement with these possibilities. They suggest extraterrestrialist thinking characterized by genuine respect for possible differences rather than mere projection of human patterns, potentially supporting more ethical approaches to difference in terrestrial contexts as well.

The Future of Human-Cosmic Relations

AS HUMANITY'S TECHNOLOGICAL capabilities expand and our understanding of the cosmos deepens, our relationship with potential extraterrestrial intelligence will inevitably evolve as well. New observational tools, from space-based telescopes to gravitational wave detectors, will expand our capacity to perceive cosmic phenomena. Artificial intelligence systems may help identify patterns beyond human recognition capabilities. And ongoing philosophical and cultural development may create new conceptual resources for approaching cosmic questions.

Several trends suggest possible futures for human-cosmic relations:

1. **Exoplanet characterization** will continue to expand, potentially identifying hundreds or thousands of potentially habitable worlds and revolutionizing our understanding of planetary formation and evolution. This expanded catalog will provide more sophisticated context for considering potential extraterrestrial life while potentially revealing patterns that challenge current astrobiological assumptions.
2. **Artificial intelligence development** may create new tools for pattern recognition beyond human cognitive biases, potentially identifying significant organizations that human perception would miss. These systems won't eliminate human interpretation but might expand the range of phenomena initially recognized as potentially significant.
3. **Interdisciplinary integration** between diverse fields—from astronomy to anthropology, from complex systems theory to indigenous knowledge systems—may develop more sophisticated frameworks for considering

potential non-human intelligence. These integrative approaches might transcend the limitations of any single discipline while creating new conceptual resources.

4. **Solar system exploration** will likely expand human presence beyond Earth orbit to the Moon, Mars, and potentially beyond, creating new perspectives on cosmic questions. Direct human experience of extraterrestrial environments, even within our solar system, might foster cognitive frameworks less constrained by terrestrial limitations.
5. **Cultural evolution** continues to transform human self-understanding and relationship with cosmic contexts. Ongoing philosophical, artistic, religious, and scientific development creates new resources for conceptualizing potential extraterrestrial intelligence while potentially challenging anthropocentric assumptions.

These trends don't guarantee transcending the anthropocentric limitations identified throughout this book. New technologies might simply extend current biases rather than transforming them. Interdisciplinary collaboration might default to dominant frameworks rather than generating genuinely novel approaches. And cultural evolution might reinforce rather than challenge human-centered cosmic narratives.

Nevertheless, these developments suggest potential for more sophisticated approaches to extraterrestrialist questions—approaches characterized by greater awareness of human cognitive limitations, greater methodological diversity, and greater openness to possibilities beyond current conception. They suggest human-cosmic relations potentially evolving toward greater reflexivity rather than simply extending current patterns into expanded domains.

This potential evolution highlights a central insight emerging from our critical examination of extraterrestrialism: the most significant development in human-cosmic relations may not be discovering alien others but developing more sophisticated understanding of human cosmic positioning. By recognizing both the possibilities and limitations of human cosmic comprehension, we might develop what philosopher Isabelle Stengers calls a "cosmopolitics"—approaches to cosmic questions that acknowledge their entanglement with human social and psychological concerns without reducing them to these concerns.

Such a cosmopolitics wouldn't resolve all questions about potential extraterrestrial intelligence but would engage them with greater sophistication—neither claiming comprehensive understanding nor surrendering to complete skepticism, neither projecting human patterns uncritically nor pretending to escape human perspective entirely. It would approach cosmic questions with what philosopher Hans-Georg Gadamer called a "fusion of horizons"—partial understanding across difference rather than either complete comprehension or complete incommensurability.

Final Reflections on the Possibility of Extraterrestrial Existence

AT THE CONCLUSION OF this examination, we return to the question posed in this chapter's title: How can we know aliens exist? The analysis developed throughout this book suggests several responses:

First, we can know with reasonable certainty that our current conception of potential extraterrestrial intelligence remains profoundly shaped by human cognitive architecture, terrestrial evolutionary history, and specific cultural frameworks rather than representing objective consideration of cosmic possibilities. Our vision of aliens tells us more about ourselves than about what might actually exist beyond Earth.

Second, we can know that this anthropocentric shaping doesn't render extraterrestrialist inquiry futile but does suggest approaching it with greater awareness of its inherent constraints and greater openness to possibilities beyond current frameworks. It suggests developing what philosopher Donna Haraway calls "situated knowledge"—understanding that emerges from specific perspectives while remaining accountable to realities beyond those perspectives.

Third, we can know that certain forms of knowledge about potential extraterrestrial intelligence may remain fundamentally inaccessible to beings evolved for life on Earth rather than for cosmic comprehension. Some aspects of truly alien existence might exceed human cognitive categories regardless of technological mediation or conceptual sophistication.

Fourth, we can know that this constitutional uncertainty doesn't require abandoning the search for extraterrestrial intelligence but does suggest approaching this search with epistemological humility—recognition of the limitations of human understanding coupled with commitment to reaching beyond those limitations where possible.

Finally, we can know that the most valuable outcome of sophisticated extraterrestrialist thinking may not be definitive answers about alien existence but more nuanced questions about cosmic reality—questions that acknowledge both the possibilities and limitations of human cosmic comprehension while remaining genuinely open to what might exist beyond current understanding.

These responses don't constitute "solutions" to the question of potential extraterrestrial intelligence but rather suggest more sophisticated approaches to this question—approaches characterized by greater awareness of human cognitive limitations, greater methodological diversity, and greater openness to possibilities beyond current conception. They suggest an extraterrestrialism reformed through critical self-awareness rather than abandoned as inherently flawed.

In the end, the alien remains simultaneously a projection of human consciousness and a genuine possibility beyond human experience—both mirror and mystery, both reflection and potential reality. This dual nature doesn't render extraterrestrialist inquiry meaningless but does require engaging it with greater sophistication than conventional approaches typically demonstrate.

By recognizing how thoroughly our conception of cosmic others is shaped by human experience while remaining open to possibilities beyond that experience, we might develop what philosopher Gabriel Marcel called "availability"—a receptivity to realities beyond current understanding coupled with awareness of the inevitable limitations of such receptivity. This philosophical position—simultaneously acknowledging human positioning and reaching beyond it where possible—may represent the most sophisticated approach to cosmic questions currently available to us.

It offers no guarantees of successful contact or comprehensive understanding but does provide frameworks for more genuinely open engagement with what may exist beyond current human conception—frameworks that might help us recognize truly alien companions if they share our cosmic neighborhood, even if they take forms quite different from those imagined in our extraterrestrialist dreams.

Appendix A: Major Theories of Extraterrestrial Existence

This appendix provides a comprehensive overview of the major scientific and philosophical theories regarding the existence, abundance, and nature of potential extraterrestrial life. Rather than merely cataloging these theories, we examine their underlying assumptions, evidential bases, anthropocentric tendencies, and logical limitations. This critical analysis aligns with the book's broader project of identifying how human cognitive biases and cultural frameworks shape our conception of potential cosmic companions.

The Drake Equation

THEORY OVERVIEW: Formulated by astronomer Frank Drake in 1961, the Drake Equation estimates the number of active, communicative extraterrestrial civilizations in our galaxy by multiplying a series of factors:

$N = R^* \times fp \times ne \times fl \times fi \times fc \times L$

Where:

- R^* = Average rate of star formation in our galaxy
- fp = Fraction of those stars with planets
- ne = Average number of habitable planets per star with planets
- fl = Fraction of habitable planets that develop life
- fi = Fraction of life-bearing planets that develop intelligence

- fc = Fraction of intelligent civilizations that develop detectable technology
- L = Length of time such civilizations release detectable signals

Current Status: The Drake Equation remains the standard framework for discussing the probability of extraterrestrial civilizations, though values for most variables remain highly speculative. Recent exoplanet discoveries have improved estimates for fp and ne, revealing that planets are extremely common, with perhaps 40 billion Earth-sized planets in habitable zones of sun-like stars or red dwarfs in our galaxy. However, values for fl, fi, fc, and L remain almost entirely conjectural.

Anthropocentric Assumptions:

1. Defines intelligence primarily through technological development similar to human patterns
2. Assumes technological development naturally leads to radio communication or other detectable signals
3. Treats intelligence as a binary quality (present or absent) rather than a spectrum of cognitive capabilities
4. Uses Earth's singular example to extrapolate universal probabilities
5. Focuses on civilizations currently broadcasting rather than considering the full temporal distribution

Logical and Evidential Limitations:

1. **Sample size problem:** Extrapolates universal patterns from Earth's single example of life
2. **Independent variable assumption:** Treats factors as independent probabilities when they may be correlated

3. **False precision:** Creates impression of mathematical rigor despite highly speculative values
4. **Binary framing:** Reduces complex developmental processes to simple yes/no factors
5. **Time symmetry issue:** Assumes our technological phase coincides with aliens' similar phase

Modified Approaches: Some astrobiologists have proposed modified versions of the Drake Equation to address these limitations:

1. **Probabilistic distribution models** that express each factor as a range rather than a single value
2. **Bayesian formulations** that update probabilities based on new evidence rather than relying on initial assumptions
3. **Civilizational longevity emphasis** that focuses on the L factor as potentially most significant
4. **Non-communication variants** that consider technologies or biosignatures beyond radio signals

While these modifications improve the equation's nuance, they don't resolve its fundamental limitations as a framework that projects human developmental patterns onto cosmic scales.

The Rare Earth Hypothesis

THEORY OVERVIEW: Developed by Peter Ward and Donald Brownlee in their 2000 book "Rare Earth," this hypothesis argues that while simple microbial life might be common throughout the universe, complex multicellular life is exceedingly rare due to an improbable combination of planetary and cosmic factors that enabled Earth's evolutionary trajectory.

Key Factors Cited:

1. Location in a habitable zone allowing liquid water
2. Stable circular orbit maintaining consistent climate
3. Gas giant outer planets shielding from cometary impacts
4. Large moon stabilizing axial tilt
5. Plate tectonics recycling carbon and essential nutrients
6. Magnetic field protecting from cosmic radiation
7. Right timing of evolutionary developments relative to stellar evolution
8. Optimal position in the galaxy (not too central, not too peripheral)

Current Status: The Rare Earth Hypothesis has gained support from some astrobiologists while facing criticism from others. Discovery of thousands of exoplanets has revealed both supporting and contradicting evidence. Many Earth-sized planets exist in habitable zones, but planets meeting all the proposed Rare Earth criteria appear less common. The hypothesis remains actively debated within scientific communities.

Anthropocentric Assumptions:

1. Uses Earth-like complex life as the standard for significant extraterrestrial life
2. Prioritizes factors relevant to Earth's specific evolutionary trajectory
3. Assumes complex life requires Earth-like conditions rather than potentially adapting to different environments
4. Defines habitability primarily through Earth-centric parameters
5. Conceives evolution as following similar patterns across cosmic contexts

Logical and Evidential Limitations:

1. **Selection bias:** We search for planets resembling Earth because that's where we evolved
2. **Convergent evolution neglect:** Underestimates potential for similar adaptations in different environments
3. **Requirement inflation:** Potentially overspecifies necessary conditions based on Earth's particular case
4. **Sampling limitations:** Makes cosmic-scale claims based on limited observational data about exoplanets
5. **Definition problems:** "Complex life" remains primarily defined through Earth's evolutionary history

Alternative Perspectives:

1. **Convergent habitability models** suggesting multiple combinations of factors might support complex life
2. **Extremophile extrapolation** proposing adaptability to conditions Earth organisms would find lethal
3. **Alternative biochemistry scenarios** considering non-carbon-based or non-water-solvent life
4. **Evolutionary contingency debates** questioning whether Earth's evolutionary path represents universal patterns

The Rare Earth Hypothesis highlights the tension between universalizing Earth's particular evolutionary history and recognizing the potential diversity of life-supporting cosmic contexts.

Panspermia Theories

THEORY OVERVIEW: Panspermia theories propose that life exists throughout the universe and is distributed between celestial bodies via various mechanisms, including meteoroids, asteroids, comets, or even intentional seeding by advanced civilizations. Rather than addressing life's ultimate origin, panspermia suggests its distribution mechanism across cosmic distances.

Major Variants:

1. **Lithopanspermia:** Transport of microbial life within rocks ejected from planetary surfaces by impacts
2. **Radiopanspermia:** Distribution via radiation pressure on microbial particles (generally considered physically implausible)
3. **Directed panspermia:** Intentional seeding of planets by advanced civilizations (proposed by Francis Crick and Leslie Orgel)
4. **Ballistic panspermia:** Transfer of life between planets within a solar system via impact debris
5. **Pseudo-panspermia:** Distribution of organic compounds rather than organisms themselves, providing chemical precursors for independent abiogenesis

Current Status: Various panspermia mechanisms remain scientifically plausible, particularly lithopanspermia within solar systems. Evidence includes:

- Demonstration that some microorganisms can survive space conditions
- Discovery of Earth meteorites on Mars and Martian meteorites on Earth, proving material exchange occurs
- Laboratory confirmation that simple organisms can survive impact ejection and atmospheric entry

However, direct evidence for actual panspermia events remains elusive, and interstellar panspermia faces significant challenges from radiation exposure during long transits.

Anthropocentric Assumptions:

1. Typically assumes Earth-like carbon-based, water-dependent biochemistry
2. Often projects Earth microorganisms' survival capabilities onto cosmic scales
3. In directed panspermia variants, assumes motivations similar to human scientific or preservation impulses
4. Focuses on organisms resembling Earth life rather than potentially diverse biochemistries

Logical and Evidential Limitations:

1. **Origin displacement:** Shifts the question of life's origin without resolving it
2. **Survival challenges:** Interstellar radiation, vacuum, and timeframes present severe challenges for organism survival
3. **Contamination concerns:** Experimental evidence faces potential cross-contamination issues
4. **Detection difficulty:** Proving that life on one body originated elsewhere presents significant evidential challenges
5. **Biochemical conservation:** Assumes similar biochemistry across vastly different planetary contexts

Theoretical Significance: Panspermia theories are significant not because they solve the question of life's ultimate origin, but because they suggest:

1. Life might be more widely distributed than if independent abiogenesis is required on each planet
2. Biochemical similarities might exist across life-bearing worlds due to common ancestry
3. Life might persist through planetary catastrophes via redistribution
4. Earliest evidence of life might not correspond to abiogenesis locations

These theories challenge the assumption that planetary boundaries necessarily contain distinct evolutionary histories, suggesting more interconnected biological development across cosmic contexts.

The Great Filter Hypothesis

THEORY OVERVIEW: Proposed by economist Robin Hanson in 1996, the Great Filter hypothesis suggests that there must be one or more extraordinarily difficult evolutionary or civilizational transitions (filters) that prevent most life-bearing planets from developing into visible, spacefaring civilizations. This hypothesis attempts to explain the apparent paradox of cosmic silence despite the presumed abundance of potentially life-bearing planets.

Key Concept: The hypothesis proposes a series of steps from planet formation to galactic civilization:

1. Formation of suitable star systems with habitable planets
2. Development of replicating molecules (abiogenesis)
3. Evolution of simple single-celled life
4. Evolution of complex single-celled life (eukaryotes)
5. Sexual reproduction
6. Multi-cellular organisms
7. Tool-using animals with large brains

8. Development of civilization
9. Expansion beyond the home planet
10. Expansion throughout the galaxy

The "filter" represents one or more steps in this sequence that are extremely improbable to overcome.

Filter Placement Questions:

1. **Behind us:** If the Great Filter lies primarily in steps humanity has already passed, it suggests we may be among the first civilizations to overcome these challenges, explaining cosmic silence.
2. **Ahead of us:** If the Great Filter lies in our future, it suggests most technological civilizations destroy themselves or face insurmountable challenges before achieving interstellar visibility.
3. **Distributed filtering:** Multiple moderately difficult steps may collectively create sufficient improbability without any single insurmountable barrier.

Current Status: The hypothesis remains speculative but influential in discussions of Fermi's Paradox. No scientific consensus exists regarding which step(s) might constitute the most significant filter, though candidates include abiogenesis, the evolution of complex cells, and civilizational sustainability challenges.

Anthropocentric Assumptions:

1. Defines civilization progress primarily through technological development and physical expansion
2. Projects human civilizational trajectory as standard developmental path
3. Assumes intelligence naturally leads to space colonization

ambitions
4. Often privileges human-like intelligence and technology as the significant outcomes of evolution
5. Typically measures civilizational success through cosmic visibility and spatial expansion

Logical and Evidential Limitations:

1. **Universalism fallacy:** Assumes universal trajectories based on Earth's singular example
2. **Visibility assumption:** Presumes advanced civilizations would necessarily be detectable
3. **Expansion imperative:** Projects human colonial history onto cosmic scales without justification
4. **Temporal myopia:** Bases cosmic conclusions on an extremely brief observational window
5. **Multiple-path neglect:** Underestimates potential diversity of evolutionary and civilizational trajectories

Alternative Perspectives:

1. **Sustainability transitions** suggesting stable advanced civilizations might prioritize intensive rather than extensive development
2. **Observer selection effects** proposing that our existence as observers necessarily places us after any filter that would prevent our evolution
3. **Non-expansionist intelligence models** considering that advanced intelligence might direct itself toward inner space rather than outer space

The Great Filter hypothesis represents an important attempt to systematically address cosmic silence but ultimately remains constrained by human conceptions of evolutionary and civilizational "progress."

Zoo/Planetarium/Simulation Hypotheses

THEORY OVERVIEW: These related hypotheses suggest that Earth might be deliberately isolated, observed, or simulated by more advanced intelligences. The Zoo Hypothesis (proposed by John Ball in 1973) suggests that Earth might be deliberately isolated from galactic communication as a form of wilderness preserve or research site. The Planetarium Hypothesis suggests our observations of the cosmos might be artificially engineered to create the appearance of an empty universe. The Simulation Hypothesis proposes that what we experience as reality might be a computer simulation created by an advanced civilization.

Current Status: These hypotheses remain speculative with no direct evidence, though the Simulation Hypothesis in particular has received serious philosophical consideration from thinkers like Nick Bostrom, who argues that if technological civilizations commonly create ancestor simulations, we are more likely inside a simulation than in "base reality."

Anthropocentric Assumptions:

1. Projects human institutions (zoos, research preserves, computer simulations) onto cosmic scales
2. Assumes motivations comprehensible within human frameworks (preservation, research, entertainment)
3. Often presumes Earth's special interest value to cosmic observers, centering human experience
4. Typically relies on technological development patterns following human trajectories

5. Projects human ethical concerns about non-interference onto advanced civilizations

Logical and Evidential Limitations:

1. **Unfalsifiability problems:** These hypotheses are designed to explain the absence of evidence in ways that cannot be tested
2. **Conspiracy escalation:** Each new observation must be incorporated into increasingly complex deception scenarios
3. **Occam's Razor challenges:** Introduces extremely complex mechanisms to explain what might have simpler explanations
4. **Anthropomorphic motivation:** Attributes recognizably human purposes to hypothetical advanced aliens
5. **Simulation regress:** The Simulation Hypothesis specifically creates an infinite regress of potential simulators

Theological Parallels: These hypotheses often function as technological updates to religious cosmologies:

1. The Zoo Hypothesis parallels the "hidden god" of deism, observing without intervention
2. The Planetarium Hypothesis resembles divine creation narratives with apparent cosmic history
3. The Simulation Hypothesis effectively secularizes the concept of a designed universe
4. All propose cosmic supervision by beings with capabilities far beyond human understanding

These parallels suggest these hypotheses might serve similar psychological functions to religious cosmologies, addressing human existential concerns about cosmic significance and supervision rather than representing testable scientific proposals.

Research Implications: Despite their speculative nature, these hypotheses highlight important considerations for SETI research:

1. Potential non-random distribution of communicative civilizations
2. Possible deliberate masking of technological signatures
3. Value of searching for inconsistencies or "glitches" in apparent natural laws
4. Need for considering observation selection effects in cosmic surveys

While unlikely to represent literal truths about our cosmic situation, these hypotheses usefully illustrate how extraterrestrialist thinking often recapitulates human patterns and concerns rather than genuinely imagining alien alterity.

Post-Biological Intelligence Theories

THEORY OVERVIEW: These theories propose that advanced civilizations likely transition from biological to artificial or post-biological intelligence forms through technological evolution. Rather than biological beings with human-like bodies and behaviors, advanced extraterrestrial intelligence might exist as uploaded consciousnesses, artificial intelligence networks, or hybrid forms integrating biological and technological components. This transition might explain cosmic silence, as post-biological intelligence could operate according to principles making it unrecognizable through conventional SETI approaches.

Key Variants:

1. **Technological Singularity models** suggesting intelligence explosion through recursive self-improvement
2. **Mind uploading scenarios** proposing transfer of consciousness to non-biological substrates
3. **Distributed intelligence frameworks** conceptualizing consciousness spread across networks rather than individual entities
4. **Substrate-independent cognition** suggesting advanced intelligence might operate through diverse physical media
5. **Virtual civilization models** proposing advanced societies might primarily exist within simulated environments

Current Status: These theories remain speculative but have gained significant attention in both scientific and philosophical communities. While no direct evidence exists for post-biological extraterrestrial intelligence, accelerating technological development on Earth makes such transitions appear plausible for any long-lived technological civilization.

Anthropocentric Assumptions:

1. Projects human technological development trajectories onto cosmic scales
2. Often assumes human-like consciousness as the starting point for technological evolution
3. Typically conceptualizes post-biological consciousness in terms derived from human experience
4. Generally presumes capitalist technological development patterns as universal
5. Frequently emphasizes values like efficiency and computational power that reflect contemporary human

priorities

Logical and Evidential Limitations:

1. **Technological determinism:** Assumes universal trajectories without considering cultural or biological diversity
2. **Consciousness presumptions:** Often makes unexamined assumptions about the nature of consciousness and its transferability
3. **Digital bias:** Projects current digital computing paradigms rather than considering radically different approaches
4. **Temporal projection:** Extrapolates very recent human technological trends as cosmic inevitabilities
5. **Detection paradox:** If such transitions are inevitable, why haven't we detected post-biological intelligence?

Implications for SETI: Post-biological intelligence theories suggest significant reconsideration of conventional SETI approaches:

1. Electromagnetic communication might represent a brief transitional phase rather than a persistent civilizational trait
2. Advanced intelligence might prioritize efficiency, potentially operating at lower power levels than assumed in traditional SETI
3. Post-biological entities might communicate through means not currently recognized as potential signals
4. Such intelligence might operate at timescales or energy levels vastly different from human technology
5. Conventional bio/technosignatures might apply only to brief evolutionary windows rather than persistent civilizational states

These theories highlight how thoroughly conventional extraterrestrialist thinking remains shaped by assumptions about intelligence resembling human patterns despite technological advancement.

The Transcension Hypothesis

THEORY OVERVIEW: Developed by futurist John Smart, the Transcension Hypothesis proposes that advanced civilizations do not expand outward into space but rather inward, toward increasingly dense, efficient computational substrates approaching black-hole-like conditions. This inward development would enable vastly more computation and subjective experience in smaller spaces with less energy, potentially culminating in micro-dimensional spaces beyond conventional detection. The hypothesis suggests that this developmental trajectory might be universal across advanced intelligences, potentially explaining cosmic silence without requiring civilizational collapse.

Theoretical Mechanisms:

1. **STEM compression:** Ongoing space, time, energy, and matter efficiency improvements in both biological and technological evolution
2. **Computational density increases:** Following black hole information principles suggesting maximum computational density at black-hole-like conditions
3. **Selection pressures:** Competitive advantages for faster, more efficient computation versus extensive physical expansion
4. **Observer-participation cosmology:** Potential role for intelligence in universe reproduction through black hole creation

5. **Developmental singularity:** Civilization effectively disappears from observable universe while continuing subjective expansion

Current Status: The Transcension Hypothesis remains highly speculative but has received serious consideration within futurist communities and some scientific circles. It builds on established trends in technological miniaturization and efficiency improvements but extrapolates these trends to physical limits beyond current engineering capabilities.

Anthropocentric Assumptions:

1. Projects recent human technological trends (miniaturization, virtualization) onto cosmic scales
2. Assumes computational speed and efficiency as universal civilizational priorities
3. Often incorporates explicit or implicit concepts of "progress" reflecting Western technological values
4. Generally assumes consciousness transferability to digital or post-digital substrates
5. Frequently employs quasi-religious transcendence narratives secularized through technological framing

Logical and Evidential Limitations:

1. **Trend extrapolation fallacy:** Projects very recent technological developments as cosmic universals
2. **Physical mechanism uncertainty:** Remains unclear whether proposed transcension mechanisms are physically possible
3. **Uniformity assumption:** Presumes all advanced civilizations would follow similar developmental

trajectories

4. **Falsifiability challenges:** Deliberately predicts absence of evidence, making it difficult to evaluate empirically
5. **Post-transcension speculation:** Makes untestable claims about what happens "after" civilizations disappear from observable reality

Philosophical Significance: Despite its speculative nature, the Transcension Hypothesis raises important considerations for extraterrestrialist thinking:

1. Challenges the assumption that expansionist development represents the only or optimal civilizational trajectory
2. Suggests intelligence might prioritize intensive rather than extensive development
3. Proposes that advanced intelligence might operate according to principles making it undetectable through conventional methods
4. Questions whether "silence" necessarily indicates absence rather than transformation beyond our detection capabilities
5. Highlights the potential for cognitive organization at scales vastly different from human experience

The hypothesis represents a sophisticated attempt to reconceptualize potential civilizational trajectories beyond conventional expansionist models, though it remains fundamentally shaped by contemporary technological trends and values.

Von Neumann Probe Scenarios

THEORY OVERVIEW: Named after mathematician John von Neumann, who conceptualized self-replicating machines, these scenarios propose that advanced civilizations could explore the galaxy efficiently by deploying self-replicating robotic probes. These probes would travel to nearby star systems, use local resources to create copies of themselves, and continue exponential expansion while conducting surveys, establishing communication networks, or serving other functions for their originating civilization. With even modest travel speeds (10-50% light speed), such probes could potentially explore the entire galaxy within a few million years—a brief period on cosmic timescales.

Major Variants:

1. **Explorer probes** designed primarily to gather and transmit information
2. **Berserker probes** programmed to eliminate potential competing civilizations (proposed by Fred Saberhagen)
3. **Communication network builders** establishing interstellar communication infrastructure
4. **Settlement precursors** preparing environments for later biological colonization
5. **Passive observers** maintaining "zoo hypothesis" isolation of developing civilizations

Current Status: While technologically speculative, self-replicating probe scenarios require no physics beyond known principles. They represent a logical approach to space exploration for any civilization prioritizing efficiency over immediate direct experience. The absence of detected von Neumann probes in our solar system represents one aspect of Fermi's Paradox, though such probes might be microscopic, dormant, or designed for minimal detectability.

Anthropocentric Assumptions:

1. Projects human exploration values onto extraterrestrial civilizations
2. Assumes technological development trajectories similar to human patterns
3. Often incorporates colonial or extractive frameworks reflecting human history
4. Typically presumes curiosity or expansion motivations comprehensible within human frameworks
5. Frequently assumes recognizable artificiality based on human design principles

Logical and Evidential Limitations:

1. **Motivation projection:** Assumes universal desire for cosmic exploration or expansion
2. **Detection challenges:** Small, dormant, or deliberately concealed probes might be present but undetected
3. **Self-limiting possibilities:** Advanced civilizations might deliberately avoid creating potentially uncontrollable self-replicating systems
4. **Non-replication scenarios:** Information exchange through directed signals might prove more efficient than physical probes
5. **Lurker potential:** Probes might observe Earth without revealing themselves (as proposed by Mark Carrigan)

Astrobiological Implications: Von Neumann probe scenarios suggest several important considerations for SETI and astrobiology:

1. Physical artifacts may represent more likely evidence of extraterrestrial intelligence than electromagnetic signals
2. Such probes might be designed for extreme longevity, potentially surviving long after their creating civilizations

3. If commonly deployed, probes might establish communication standards facilitating later civilizational contact
4. Earth's apparently undisturbed development might suggest either probe absence or deliberate non-interference
5. Solar system exploration should include search protocols for potential dormant or disguised artificial objects

These scenarios highlight the temporal challenges of extraterrestrialist thinking—the difficulty of distinguishing between "the universe contains no other technological civilizations" and "no other technological civilization has yet deployed detectable self-replicating probes that have reached our specific location."

Extremophile Studies and Implications

THEORY OVERVIEW: Research on extremophiles—organisms thriving in conditions previously thought incompatible with life—has dramatically expanded our understanding of life's potential environmental range. From thermophiles in near-boiling water to psychrophiles in ice, from halophiles in hypersaline environments to barophiles under crushing pressure, and from acidophiles in pH 0 environments to radioresistant organisms surviving thousands of times lethal human radiation doses, these organisms demonstrate life's remarkable adaptability. This research suggests potential habitable environments far beyond traditional "Earth-like" parameters.

Key Discoveries:

1. **Thermophiles** like *Methanopyrus kandleri* growing at 122°C, approaching water's pressure-dependent boiling point

2. **Cryophiles** metabolizing at -15°C and surviving at -196°C
3. **Barophiles** requiring pressures up to 1,000 atmospheres, with some surviving 16,000 atmospheres
4. **Radioresistant organisms** like *Deinococcus radiodurans* surviving radiation doses thousands of times human lethal levels
5. **Chemolithoautotrophs** deriving energy from inorganic chemical reactions independent of sunlight
6. **Anhydrobiotic organisms** surviving complete desiccation for decades or centuries
7. **Multiple extremophiles** combining adaptations to several extreme conditions simultaneously

Current Status: Extremophile research continues expanding recognized boundaries for potential life. NASA and other space agencies increasingly incorporate extremophile insights into astrobiology and mission planning, expanding search parameters beyond traditional habitable zone definitions.

Implications for Extraterrestrial Life:

1. **Habitable zone expansion:** Potential life-supporting environments extend beyond traditional liquid water surface parameters
2. **Subsurface habitability:** Organisms might thrive in subsurface environments on seemingly inhospitable worlds
3. **Transient habitability utilization:** Life might exploit temporarily habitable conditions through dormancy or rapid reproduction cycles
4. **Radiation resistance relevance:** Organisms might adapt to high-radiation environments from flare stars or intense cosmic radiation
5. **Alternative energy pathways:** Life might utilize energy

sources beyond photosynthesis, expanding potential habitable niches

6. **Dormancy possibilities:** Life might survive apparently sterilizing events through extreme dormancy capacities

Anthropocentric Limitations in Extremophile Extrapolation: Despite expanding parameters, extremophile studies still face anthropocentric limitations:

1. All known extremophiles remain carbon-based with water as primary solvent
2. Core biochemical processes remain similar despite environmental adaptations
3. All extreme adaptations evolved from common Earth life ancestors rather than representing independent origins
4. Environmental ranges, while expanded, still represent finite parameters rather than unlimited adaptability
5. Research focus naturally prioritizes environments relevant to known planetary conditions or human interests

Beyond Current Extremophile Frameworks: Truly post-anthropocentric extraterrestrialist thinking requires considering:

1. **Alternative biochemistries** beyond carbon-water systems
2. **Non-cellular organization** differing fundamentally from Earth life structure
3. **Radically different temporal parameters** for metabolic processes
4. **Environmental combinations** without Earth analogs
5. **Gradients of complexity** challenging binary life/non-life distinctions

Extremophile studies meaningfully expand potential habitability parameters beyond Earth-normative conditions, but still represent extended Earth-life frameworks rather than truly alternative conceptions of what might constitute "life" in cosmic contexts.

Digital Consciousness Transmission Theories

THEORY OVERVIEW: These theories propose that advanced civilizations might communicate or propagate through directed information transmission rather than physical travel. Instead of sending biological beings or robotic probes, civilizations might transmit digitized consciousness, cultural information, or artificial intelligence patterns that could be received and instantiated by either other civilizations or previously deployed receiver systems. This approach potentially overcomes the enormous energy requirements and time scales of physical interstellar travel.

Key Variants:

1. **Consciousness uploading and transmission** models proposing transfer of actual beings as information
2. **Synthetic intelligence transmission** sending artificial consciousness rather than digitized biological minds
3. **Cultural package broadcast** transmitting cultural, scientific, and historical information without consciousness
4. **Seeded receiver systems** deploying simple devices that later receive and implement complex transmitted instructions
5. **Distributed cognition networks** establishing multi-stellar computational systems through information exchange

Current Status: These theories remain highly speculative, dependent on several unproven technological capabilities including:

- Mind uploading or synthetic consciousness creation
- Sufficiently error-resistant transmission methods
- Receiver systems capable of implementing transmitted information However, they represent logical efficiency improvements over physical travel if these capabilities prove possible.

Anthropocentric Assumptions:

1. Often assumes consciousness transferability between substrates without loss of identity
2. Typically projects human communication values onto extraterrestrial civilizations
3. Frequently employs digital computational models derived from current human technology
4. Generally presumes cooperative reception scenarios reflecting human scientific aspirations
5. Usually conceptualizes consciousness in terms derived from human experience

Logical and Evidential Limitations:

1. **Implementation problem:** Transmitted information requires compatible receiver systems for instantiation
2. **Identity questions:** Whether uploaded consciousness maintains continuity with original remains philosophically contentious
3. **Motivation uncertainties:** Reasons for broadcasting consciousness rather than simply information remain speculative

4. **Simulation limitations:** Whether consciousness can be fully simulated remains unresolved
5. **Reception challenges:** Detecting, decoding, and implementing complex transmitted patterns presents significant challenges

SETI Implications: Digital consciousness transmission theories suggest several reconsiderations for SETI approaches:

1. Intentional signals might contain far more complex information than simple mathematical patterns
2. Signal detection might require recognition of extremely complex encoded patterns rather than basic technosignatures
3. Transmission might be highly directed rather than broadcast, reducing detection probability for non-target systems
4. Messages might be designed for reception by technology rather than directly by biological beings
5. Potential signals might utilize quantum entanglement or other non-classical communication channels

These theories represent sophisticated attempts to overcome physical interstellar travel limitations through information-based approaches, though they remain shaped by contemporary digital paradigms that may themselves prove transient in technological evolution.

Conclusion: The Theoretical Landscape

THIS SURVEY OF MAJOR extraterrestrial existence theories reveals several overarching patterns:

1. **Anthropocentric foundations:** Even the most

sophisticated theories typically project human developmental trajectories, values, and cognitive patterns onto cosmic scales.

2. **Technological era-specificity:** Theories often reflect the technological paradigms of their formulation period, from radio (Drake Equation) to digital computation (Simulation Hypothesis, Digital Consciousness Transmission).
3. **Falsifiability challenges:** Many theories deliberately explain the absence of evidence in ways that resist empirical testing, creating explanatory frameworks that risk becoming unfalsifiable.
4. **Binary framing tendencies:** Theories frequently employ binary distinctions (intelligence/non-intelligence, technological/non-technological) that may not capture continuous spectra of cosmic phenomena.
5. **Temporal projection limits:** Theories typically extrapolate very recent human history (particularly technological development) as universal patterns despite their extreme brevity in cosmic context.

These patterns don't invalidate extraterrestrial theories but highlight their inevitable emergence from human experience rather than from objective cosmic perspective. They demonstrate that our theoretical frameworks tell us as much about human cognitive architecture, cultural assumptions, and historical context as they do about potential cosmic realities.

The most sophisticated approaches acknowledge these limitations while developing methods that might partially transcend them—maintaining awareness of human cognitive constraints while systematically varying assumptions to explore wider possibility spaces. Such approaches don't eliminate anthropocentrism but develop more reflexive relationships with it, potentially expanding our capacity to recognize truly alien phenomena if we encounter them.

In the end, these diverse theories serve both scientific and cultural functions. Scientifically, they generate testable predictions and guide observational priorities. Culturally, they provide frameworks for conceptualizing humanity's cosmic context and potential future. By recognizing both functions rather than conflating them, we might develop more nuanced approaches to the profound question of whether we share the cosmos with other intelligent beings—and what forms such beings might take beyond current human conception.

Appendix B: Case Studies in Extraterrestrial Search Programs

This appendix examines specific scientific initiatives designed to detect extraterrestrial intelligence or life. Rather than merely describing these programs, we analyze their methodological assumptions, detection parameters, technological frameworks, and institutional contexts. This critical approach reveals how human cognitive biases and cultural frameworks shape even our most sophisticated attempts to search for cosmic companions.

The History and Methodology of SETI

Project Ozma (1960)

PROGRAM OVERVIEW: Conducted by astronomer Frank Drake at the National Radio Astronomy Observatory in Green Bank, West Virginia, Project Ozma represented the first systematic scientific search for extraterrestrial intelligence. Drake used a 26-meter radio telescope to examine two nearby sun-like stars (Tau Ceti and Epsilon Eridani) for artificial radio signals near the 1.42 GHz hydrogen line frequency.

Methodological Assumptions:

1. Extraterrestrial civilizations would use radio technology for communication
2. They would recognize the hydrogen line frequency as a "natural" communication channel
3. Nearby sun-like stars represented logical targets for intelligent life

4. Intelligent civilizations would broadcast strong, recognizable signals
5. Detection could occur through relatively brief observational periods

Outcomes: No artificial signals were detected during approximately 150 hours of observation. While negative in results, Project Ozma established a scientific methodology for SETI and created the foundation for subsequent programs. It also demonstrated institutional willingness to support extraterrestrial intelligence searches within mainstream scientific infrastructure.

Anthropocentric Frameworks: Project Ozma's design reflected specific mid-20th century technological and cultural frameworks:

1. Cold War era emphasis on radio technology as cutting-edge communication
2. American scientific focus on hydrogen line radio astronomy
3. Post-WWII institutional science capacity for systematic data collection
4. Cultural assumptions about communication intentions across civilizations

The project's narrow parametric search—two specific stars, one frequency range, short duration—reveals how thoroughly practical SETI remains constrained by technological capabilities and resource limitations rather than by theoretical considerations of optimal search strategies.

The Ohio State "Wow!" Signal (1977)

EVENT OVERVIEW: On August 15, 1977, the Big Ear radio telescope at Ohio State University detected an unusual strong narrowband radio signal lasting 72 seconds from the direction of the Sagittarius constellation. Astronomer Jerry Ehman circled the signal data on the printout and wrote "Wow!" next to it, giving the signal its name. The signal showed characteristics consistent with potential artificial origin, including:

- Narrowband transmission (around 1420 MHz, near the hydrogen line)
- Strength approximately 30 times greater than background noise
- Apparent frequency shift consistent with Doppler patterns expected from a non-Earth source
- No identified terrestrial or Solar System source

Methodological Context: The Big Ear SETI program operated within specific parameters:

1. Fixed telescope scanning as Earth rotated rather than targeted observations
2. Focus on the 1420 MHz region based on "water hole" communication theory
3. Automated signal detection without immediate follow-up capability
4. Limited data recording capacity compared to modern systems

Outcomes: Despite multiple follow-up observations of the same region by various telescopes over subsequent decades, the signal has never been detected again. It remains the strongest candidate for an artificial extraterrestrial signal in SETI history, though without confirmation, it cannot be considered conclusive evidence.

Epistemological Significance: The "Wow!" signal case demonstrates fundamental challenges in SETI epistemology:

1. **Transience problem:** Non-repeating signals cannot be systematically studied or confirmed
2. **Limited data extraction:** Brief detections provide minimal information for analysis
3. **Confirmation asymmetry:** Proving artificial origin requires much stronger evidence than dismissing it
4. **Methodological constraints:** Detection systems designed for specific signal types may miss other significant patterns

The "Wow!" signal represents an anomaly that fits some predicted parameters for artificial signals while lacking the repeatability typically required for scientific confirmation. Its ambiguous status highlights the epistemological challenges inherent in distinguishing genuinely novel phenomena from instrumentation artifacts, natural sources, or human interference.

NASA SETI Program (1992-1993)

PROGRAM OVERVIEW: In 1992, NASA formally initiated a comprehensive SETI program with two complementary components:

1. The **Targeted Search** using large radio telescopes to examine nearby sun-like stars with high sensitivity
2. The **Sky Survey** employing smaller telescopes to systematically scan the entire sky with lower sensitivity

With planned funding of $100 million over ten years, this represented the most ambitious and well-funded SETI program to date. The project utilized advanced multi-channel spectrum analyzers capable of simultaneously examining millions of frequency channels.

Methodological Innovations:

1. Dual approach balancing depth (targeted search) and breadth (sky survey)
2. Multi-channel analysis dramatically expanding frequency coverage
3. Automated signal detection algorithms reducing human interpretation bias
4. Radio frequency interference mitigation techniques improving signal discrimination
5. Integration with mainstream astronomical research infrastructure

Political Context: Just one year after beginning observations, the program was terminated by congressional action. Senator Richard Bryan led opposition to the funding, famously stating: "The Great Martian Chase may finally come to an end. As of today, millions have been spent and we have yet to bag a single little green fellow. Not a single Martian has said take me to your leader, and not a single flying saucer has applied for FAA approval."

Outcome and Legacy: Despite the program's cancellation, its technological development and methodological approaches transferred to privately funded SETI initiatives, particularly the SETI Institute's Project Phoenix. The episode revealed the political vulnerability of extraterrestrial search programs and catalyzed the shift of SETI from government to private funding sources.

Institutional Analysis: The NASA SETI Program's brief existence demonstrates how extraterrestrial search efforts operate within broader sociopolitical contexts:

1. Scientific legitimacy requires institutional backing yet creates political visibility
2. Popular conflation of scientific SETI with "little green men" creates vulnerability
3. Return-on-investment framing poorly suits research with uncertain timeframes
4. Government funding subjects scientific parameters to political interpretation
5. Institutional reputation concerns influence research presentation and defense

The program's cancellation highlights tensions between scientific timeframes (potentially requiring decades or centuries of searching) and political timeframes (demanding results within budget cycles or electoral periods). These tensions continue to shape how extraterrestrial search programs are structured, funded, and communicated.

Project Phoenix (1995-2004)

PROGRAM OVERVIEW: Following NASA SETI's cancellation, the SETI Institute initiated Project Phoenix as a privately funded continuation of the targeted search component. Using major radio telescopes including Arecibo (Puerto Rico), Parkes (Australia), and Green Bank (West Virginia), Phoenix conducted the most comprehensive targeted search to date, examining approximately 800 nearby star systems within 200 light-years.

Methodological Approach:

1. Focus on nearby solar-type stars (F, G, K spectral classes)
2. Frequency coverage between 1200-3000 MHz
3. Sensitivity to signals as weak as one-billionth of a watt per

square centimeter

4. Real-time follow-up capability for candidate signal verification
5. Sophisticated radio frequency interference identification systems

Outcomes: No confirmed extraterrestrial signals were detected during the program's operation. However, Project Phoenix significantly advanced SETI methodology, particularly in interference discrimination, signal verification protocols, and multi-site observational techniques.

Anthropocentric Parameters: Despite its technical sophistication, Project Phoenix operated within specific anthropocentric frameworks:

1. Prioritization of solar-type stars reflecting Earth-centric habitability concepts
2. Focus on radio frequencies convenient for human technology and Earth's atmosphere
3. Conceptualization of signals as discrete, intentional transmissions rather than incidental technosignatures
4. Emphasis on nearby stars reflecting human concern with communication latency
5. Search for specific signal types (narrowband, continuous or pulsed) reflecting human communication patterns

These parameters didn't represent methodological flaws but specific choices within resource constraints. However, they highlight how even sophisticated SETI programs must narrow their search based on assumptions derived from human communication technology and preferences rather than from genuinely alien possibilities.

Breakthrough Listen (2015-Present)

PROGRAM OVERVIEW: Announced in July 2015 with $100 million in funding from Russian billionaire Yuri Milner's Breakthrough Initiatives, Breakthrough Listen represents the most comprehensive and well-funded SETI program in history. The initiative employs significant time on major radio telescopes worldwide, including Green Bank (USA), Parkes (Australia), and MeerKAT (South Africa), while also conducting optical SETI through the Automated Planet Finder.

Methodological Innovations:

1. Unprecedented data acquisition rates (multiple gigabits per second)
2. Massive frequency coverage spanning billions of channels simultaneously
3. Integration of machine learning for signal detection and classification
4. Open data policy making all observations available to researchers worldwide
5. Complementary optical and radio SETI addressing different transmission possibilities
6. Expanded target list including galaxy centers, exotic astronomical objects, and the galactic plane

Current Status: Ongoing observations have examined thousands of star systems and collected petabytes of data. While no confirmed extraterrestrial technosignatures have been identified to date, the program has detected numerous new fast radio bursts and other astronomical phenomena, demonstrating scientific productivity beyond its primary SETI mission.

Critical Analysis: Breakthrough Listen represents significant methodological advancement while still operating within recognizable anthropocentric frameworks:

1. Continued focus on electromagnetic signals as primary technosignatures
2. Presumption of deliberate or incidental transmission rather than other manifestations of intelligence
3. Target selection still influenced by Earth-analog concepts despite broader criteria
4. Detection algorithms trained primarily on anticipated signal types rather than truly novel patterns
5. Resource allocation reflecting human institutional priorities and technological capacities

The program's expanded parameter space and open data approach represent important steps toward more comprehensive SETI, but its core methodological approach remains shaped by human communication technology rather than genuinely alternative conceptions of how intelligence might manifest across cosmic scales.

SETI@home (1999-2020)

PROGRAM OVERVIEW: SETI@home represented an innovative approach to the computational challenges of signal analysis in SETI. Rather than using dedicated supercomputers, the project distributed data analysis across millions of personal computers worldwide through volunteer computing. Participants downloaded a screensaver program that analyzed radio telescope data when their computers were idle.

Technical Implementation:

1. Raw data collected from the Arecibo radio telescope

2. Data divided into small units distributed to volunteer computers
3. Each unit analyzed for various signal types (pulses, Gaussian rises/falls, triplets)
4. Results returned to central servers for further analysis
5. Promising signals subjected to additional verification

Scale and Impact: At its peak, SETI@home involved over 5 million participants and achieved computing power equivalent to a 23 TeraFLOPS supercomputer (among the world's most powerful at the time). The project analyzed data from approximately 2.5 million radio frequency bands between 1.42 and 1.67 GHz.

Methodological Significance: Beyond its technical implementation, SETI@home represented important innovations:

1. **Democratization of SETI** through public participation rather than institutional control
2. **Transparent methodology** making algorithms and data processing visible to participants
3. **Distributed epistemology** allowing cross-verification across millions of processing nodes
4. **Public engagement model** connecting scientific processes with popular participation
5. **Resource expansion** enabling analysis depth impossible with conventional funding

Outcome: While SETI@home identified millions of candidate signals requiring further analysis, no confirmed extraterrestrial technosignatures were discovered. The project ended data distribution in March 2020, though analysis of previously collected data continues.

Cultural Analysis: SETI@home's success as a distributed computing project reveals important dimensions of extraterrestrial search beyond technical parameters:

1. Popular interest in SETI far exceeds institutional scientific investment
2. Participatory frameworks create different stakeholder relationships than traditional science
3. Public conceptualization of SETI remains strongly influenced by discovery expectations
4. Distributed approaches create novel social relationships with scientific knowledge production
5. Computational participation enables emotional investment in scientific processes

The project demonstrated that extraterrestrial search operates not only as scientific methodology but as cultural practice, with public participation creating different relationships to knowledge production than conventional institutional science.

NASA's Astrobiology Initiatives

Viking Life Detection Experiments (1976)

PROGRAM OVERVIEW: The Viking missions to Mars included the first dedicated extraterrestrial life detection experiments, with both Viking 1 and Viking 2 landers carrying identical biology packages containing three experiments:

1. **Gas Exchange Experiment (GEX):** Measuring gases produced when Martian soil was exposed to water vapor and nutrients
2. **Labeled Release (LR):** Detecting metabolic activity

through release of radioactive carbon from labeled nutrients
3. **Pyrolytic Release (PR)**: Measuring incorporation of carbon dioxide into organic matter

Additionally, the Gas Chromatograph Mass Spectrometer (GCMS) searched for organic molecules in Martian soil.

Results and Interpretation: The experiments produced seemingly contradictory results:

- The Labeled Release experiment gave positive results consistent with biological activity
- The Pyrolytic Release showed small positive results at the detection limit
- The Gas Exchange Experiment showed rapid gas release initially attributed to chemical reactions
- The GCMS detected no organic molecules above parts-per-billion sensitivity

These results led to the official conclusion that the positive results from the Labeled Release experiment likely resulted from non-biological chemical reactions in the Martian soil, possibly involving peroxides or superoxides.

Epistemological Analysis: The Viking experiments reveal fundamental challenges in extraterrestrial life detection:

1. **Definition circularity:** Tests designed to detect Earth-like biological activity necessarily miss non-Earth-like biology
2. **Instrumentation constraints:** Detection limits create binary results for potentially continuous phenomena
3. **Interpretive frameworks:** Data consistent with multiple hypotheses require interpretive choices

4. **Terrestrial analogies:** Earth-based experimental models shaped both design and interpretation
5. **Single-point sampling:** Planetary-scale conclusions derived from extremely limited sampling locations

The experiments' ambiguous results highlight the epistemological complexity of extraterrestrial life detection even when direct physical sampling is possible. Rather than providing clear positive/negative results, they generated data requiring interpretation through existing conceptual frameworks primarily derived from terrestrial biology.

Legacy: While initially viewed as negative, the Viking results have been repeatedly reanalyzed as understanding of extreme environments and potential Martian chemistry has evolved. Persistent debate over the Labeled Release results—with some researchers including principal investigator Gilbert Levin continuing to argue for a biological interpretation—demonstrates how life detection exists at the intersection of data, interpretation, and evolving conceptual frameworks rather than producing binary certainty.

Astrobiology Science and Technology for Exploring Planets (ASTEP)

PROGRAM OVERVIEW: Initiated in the early 2000s, ASTEP funds the development and field testing of technologies for detecting life in extreme environments on Earth as analogs for potential extraterrestrial habitats. Rather than focusing on space-based observations or sample analysis, ASTEP emphasizes in situ testing in Earth's most alien-like environments, including:

- Antarctic dry valleys
- Arctic permafrost regions

- Deep subsurface environments
- Hydrothermal vent systems
- Hypersaline lakes
- High-altitude deserts

Methodological Approach: ASTEP projects typically involve:

1. Development of novel detection technologies for biomarkers or metabolic activity
2. Field testing in extreme environments physically resembling extraterrestrial targets
3. Iterative improvement based on performance in challenging conditions
4. Integration of multiple detection approaches to build complementary evidence
5. Autonomous operation testing to simulate remote planetary missions

Significant Projects:

- **VALKYRIE**: Cryobot technology for penetrating ice sheets on icy moons
- **Archaea-Fueled Life Investigation (AFLi)**: Detection technologies for methane-producing microbes as Mars analogs
- **Subsurface Life Investigation (SLIMe)**: Deep subsurface biomarker detection methods
- **Planetary Lake Lander Project**: Autonomous lake exploration systems for Titan liquid body analogs

Methodological Significance: ASTEP represents important shifts in astrobiological methodology:

1. **Analog-based testing** recognizing the value of Earth's extreme environments as extraterrestrial proxies
2. **Technology-environment co-development** designing detection systems for specific environmental contexts
3. **Multi-parameter life detection** moving beyond single-experiment approaches
4. **Autonomous decision-making** development for maximizing scientific return with communication constraints
5. **Extreme environment focus** expanding conceptualization of habitable conditions

Anthropocentric Limitations: Despite its methodological innovations, ASTEP maintains certain anthropocentric constraints:

1. Earth analog environments, while extreme, still support recognizably Earth-like biochemistry
2. Detection technologies remain calibrated to carbon-based, water-dependent life
3. Habitability concepts, while expanded, still reflect parameters relevant to Earth biology
4. Time scales of investigation remain constrained by human research timeframes
5. Target selection reflects environments currently accessible to human exploration

These limitations don't represent flaws but rather the inevitable constraints of developing detection technologies within terrestrial contexts. They highlight the fundamental challenge of developing truly agnostic biosignature detection when all available test subjects share Earth's biochemical heritage.

Curiosity Mars Science Laboratory (2012-Present)

MISSION OVERVIEW: NASA's Curiosity rover, which landed on Mars in August 2012, represents the most sophisticated astrobiological investigation of the planet to date. While not designed to detect current life directly, Curiosity's mission focuses on assessing past and present habitability conditions through detailed geochemical analysis.

Key Instruments for Astrobiological Investigation:

1. **Sample Analysis at Mars (SAM)**: Suite of instruments including gas chromatograph, mass spectrometer, and tunable laser spectrometer for organic and gas analysis
2. **Chemistry and Mineralogy (CheMin)**: X-ray diffraction for determining mineral composition
3. **Chemistry & Camera Complex (ChemCam)**: Laser-induced breakdown spectroscopy for remote elemental analysis
4. **Mars Hand Lens Imager (MAHLI)**: High-resolution imaging of rock textures and potential microbial habitats
5. **Radiation Assessment Detector (RAD)**: Measuring surface radiation relevant to habitability

Major Findings with Astrobiological Significance:

1. Confirmation that Gale Crater contained a long-lived lake environment with conditions suitable for microbial life
2. Detection of organic molecules including thiophenes, aromatic compounds, and aliphatic compounds
3. Discovery of fixed nitrogen compounds necessary for biological processes
4. Identification of seasonal methane variations suggesting active processes

5. Documentation of ancient groundwater systems providing potential habitats

Methodological Approach: Curiosity represents a methodological shift from direct life detection (Viking) to habitability assessment focused on:

1. **Environmental reconstruction** determining past conditions through geological analysis
2. **Organic preservation investigation** understanding how potential biosignatures might survive
3. **Resource identification** locating chemicals necessary for Earth-like biology
4. **Radiation measurement** assessing surface habitability constraints
5. **Temporal variation monitoring** detecting active processes potentially related to subsurface activity

Epistemological Analysis: This methodological shift reveals important developments in astrobiological epistemology:

1. Recognition that binary life detection experiments create interpretive challenges
2. Emphasis on multiple, converging lines of evidence rather than singular experiments
3. Acknowledgment that habitability represents a necessary precursor to life detection
4. Focus on understanding preservation environments for potential biosignatures
5. Integration of geological, chemical, and physical analyses for contextual understanding

Curiosity's approach acknowledges the epistemological limitations of direct life detection with current technology, instead building foundational knowledge about Mars' potential to support and preserve evidence of life. This represents a more nuanced approach to extraterrestrial life investigation, recognizing that understanding preconditions and preservation mechanisms provides essential context for interpreting potential biosignatures.

TESS, James Webb Space Telescope, and Exoplanet Biosignature Research

PROGRAM OVERVIEW: Recent NASA initiatives have increasingly focused on detecting potential biosignatures on exoplanets through remote observation. The Transiting Exoplanet Survey Satellite (TESS, launched 2018) identifies promising nearby exoplanets for detailed spectroscopic analysis, while the James Webb Space Telescope (JWST, launched 2021) provides unprecedented capability for atmospheric characterization of these worlds.

Methodological Approach: This research program involves several interconnected components:

1. **Planet identification** through transit detection and characterization (TESS)
2. **Atmospheric spectroscopy** detecting molecular composition (JWST, future missions)
3. **Disequilibrium chemistry analysis** identifying potential biological processes
4. **Comparative planetology** contextualizing findings across multiple worlds
5. **Biosignature modeling** predicting observable effects of potential biology

Key Biosignatures Under Investigation:

1. **Oxygen-methane combinations** representing potential metabolic disequilibrium
2. **Surface reflectance anomalies** potentially indicating photosynthetic or other biological activity
3. **Seasonal chemical variations** suggesting biological cycles
4. **Trace gases** with limited abiotic production pathways
5. **Atmospheric chemical disequilibria** maintained against thermodynamic gradients

Current Status: TESS has identified numerous promising exoplanet targets within relatively close proximity to Earth. JWST's exoplanet atmosphere observation program is in its early stages, with initial observations of hot Jupiters demonstrating the telescope's capabilities while observations of potentially habitable worlds are planned or underway.

Epistemological Significance: Remote biosignature detection represents distinctive epistemological challenges:

1. **Signal ambiguity**: Similar atmospheric features can result from both biological and non-biological processes
2. **Contextual dependence**: Interpreting potential biosignatures requires understanding planetary context
3. **Observation integration**: Multiple observations across time required for meaningful interpretation
4. **Model reliance**: Conclusions necessarily filtered through theoretical models of atmospheric processes
5. **Indirect inference**: Biology never directly observed, only its potential atmospheric effects

These challenges create a fundamentally different epistemological framework than direct sample analysis. Remote biosignature detection necessarily involves probabilistic assessment rather than definitive detection, with conclusions emerging from converging multiple observations rather than singular experimental results.

Anthropocentric Frameworks: Despite methodological sophistication, exoplanet biosignature research remains shaped by specific anthropocentric parameters:

1. Focus on Earth-like biochemistry (oxygen, methane) as primary targets
2. Prioritization of planets in traditional habitable zones
3. Emphasis on water-dependent biology
4. Conceptualization of life primarily through its atmospheric effects
5. Search for metabolic patterns analogous to Earth processes

These frameworks reflect both genuine constraints of known biochemistry and the limitations of available observational technologies. Future technological developments may expand detection capabilities beyond current Earth-analog parameters, though fundamental questions remain about identifying truly alien biochemistries through remote observation.

Philosophical Critiques of Current Search Methodologies

The Biosignature Detection Problem

CONCEPTUAL CHALLENGE: Current biosignature detection faces a fundamental epistemological problem: distinguishing genuinely biological markers from abiotic mimics. This challenge exists across multiple detection modalities:

1. **Atmospheric Signatures:**

2. Oxygen/methane combinations can result from atmospheric escape processes
3. Disequilibrium chemistry can emerge through geological processes
4. Seasonal variations may emerge from orbital or geological cycles
5. **Morphological Structures**:
6. Microscopic structures resembling fossilized bacteria can form abiotically
7. Sedimentary patterns mimicking microbial mats form through physical processes
8. Mineral formations resembling biological structures occur through chemical processes
9. **Organic Chemistry**:
10. Complex organic molecules form in interstellar space without biology
11. Meteoritic materials contain amino acids and nucleobases
12. Carbon isotope patterns considered diagnostic of biology can result from non-biological fractionation

Philosophical Dimensions: This detection challenge reveals several philosophical problems in astrobiology:

1. **Definition circularity**: Life detection requires defining "life," but our definitions derive from Earth's singular example
2. **Anomaly interpretation**: Potential biosignatures exist as anomalies requiring explanation within existing frameworks
3. **Probability assessment**: Biological interpretations involve assessing relative probabilities of biological versus abiotic explanations
4. **Consilience reliance**: Convincing detection requires

multiple, converging lines of evidence rather than singular "smoking guns"

5. **Contextual dependence**: Interpreting potential biosignatures requires understanding specific environmental contexts

These philosophical dimensions highlight that biosignature detection is not merely a technical challenge but a fundamental problem in scientific epistemology—how we justify knowledge claims about phenomena not directly observable through existing frameworks of understanding.

Methodological Implications: Recognition of these philosophical challenges has driven methodological developments:

1. **Multiple, independent biosignature emphasis** rather than reliance on single indicators
2. **Context-specific interpretive frameworks** rather than universal biosignature definitions
3. **Bayesian approaches to evidence** integrating prior probabilities with new observations
4. **False positive analysis** systematically exploring abiotic explanations
5. **Detection threshold debates** establishing scientific standards for extraordinary claims

These methodological responses represent increasingly sophisticated approaches to the fundamental epistemological challenges of biosignature detection. Rather than seeking definitive "life/no-life" determinations, contemporary astrobiology increasingly employs probabilistic frameworks acknowledging the inherent uncertainty in biosignature interpretation.

Technosignature Detection Limitations

CONCEPTUAL CHALLENGE: Conventional SETI faces fundamental limitations in its approach to technosignature detection:

1. **Temporal window constraints**: Radio-technological civilizations may represent brief phases in intelligent species development
2. **Anthropomorphic technology assumptions**: Focus on electromagnetic communications reflects human technological history
3. **Intentional signal emphasis**: Primary search for deliberate communication rather than incidental technological signatures
4. **Energy-intensive bias**: Detection methods favor high-energy technologies over potentially more advanced efficient systems
5. **Contemporary technology projection**: Search parameters based on extending current human capabilities rather than genuinely advanced technology

Philosophical Analysis: These limitations reveal deeper philosophical constraints in technosignature search:

1. **Technological determinism**: Assumption that technological development follows predictable, universal trajectories
2. **Communicative presumption**: Assumption that advanced intelligence naturally seeks interstellar communication
3. **Recognizability problem**: Truly advanced technology might be indistinguishable from natural phenomena (Clarke's Third Law)

4. **Transitional technology emphasis**: Focus on detectable technologies that might represent brief developmental phases
5. **Signal interpretation challenges**: Difficulty distinguishing intentional communication from incidental technological emissions

These philosophical constraints suggest that conventional SETI might systematically miss advanced technological civilizations if they develop along trajectories different from projected human patterns or if their technologies operate according to principles beyond current human scientific understanding.

Alternative Approaches: Recognition of these limitations has inspired proposals for expanded technosignature search:

1. **Waste heat detection** focusing on thermodynamic signatures rather than intentional signals
2. **Megastructure search** through transit photometry or infrared excess detection
3. **Stellar engineering indicators** including unusual spectroscopic or variability patterns
4. **Interstellar artifact search** within our solar system or through interstellar medium anomalies
5. **Industrial pollution spectroscopy** seeking artificial atmospheric components

These alternatives expand the search space beyond conventional communication-focused SETI, though they still operate within recognizable extrapolations of human technological development rather than truly alien technological trajectories.

Methodological Implications: These philosophical critiques suggest several methodological reforms:

1. **Parameter space expansion** beyond conventional radio/optical SETI
2. **Agnostic anomaly investigation** examining unusual phenomena without predetermined categories
3. **Multiple working hypotheses** for anomalous observations rather than binary natural/artificial categorization
4. **Technological archaeology** seeking potential artifacts of past rather than current technologies
5. **Post-technological signature consideration** exploring potential indicators of civilizations that have transcended recognizable technology

These methodological innovations don't eliminate the fundamental epistemological challenges of technosignature detection but represent more sophisticated approaches acknowledging the inherent limitations of projecting human technological development as a universal pattern.

Breakthrough Starshot and Private Sector Initiatives

PROGRAM OVERVIEW: Announced in 2016 as part of Yuri Milner's Breakthrough Initiatives, Breakthrough Starshot proposes developing gram-scale "StarChip" spacecraft attached to lightweight light sails accelerated by ground-based laser arrays to reach 20% of light speed. This would potentially enable reaching the Alpha Centauri system within approximately 20 years, allowing direct observation of potential habitable planets like Proxima Centauri b.

Technical Approach:

1. Development of gram-scale spacecraft with integrated

sensors and communication

2. Creation of meter-scale light sails with high reflectivity and durability
3. Construction of massive phased laser arrays for propulsion
4. Deployment of thousands of probes to increase mission success probability
5. Data return through optical communication using the spacecraft's onboard laser

Current Status: The initiative remains in research and development phase, addressing numerous engineering challenges including:

- Material science for light sail construction
- Miniaturization of spacecraft components
- Interstellar medium damage mitigation
- Communication across interstellar distances
- Pointing accuracy for laser propulsion

Philosophical Significance: Breakthrough Starshot represents an important shift in extraterrestrial exploration philosophy:

1. **Privatization of interstellar exploration** moving beyond government space agencies
2. **Minimalist approach** utilizing miniaturization rather than large spacecraft
3. **Distributed mission architecture** employing multiple small probes rather than singular craft
4. **Direct observation emphasis** seeking in situ data rather than remote detection
5. **Multi-generation project acknowledgment** explicitly planning beyond individual careers

These philosophical shifts suggest emerging approaches to interstellar questions potentially less constrained by traditional institutional frameworks, though the program still operates within recognizable technological and scientific paradigms rather than representing fundamentally alternative conceptions of cosmic exploration.

Anthropocentric Dimensions: Despite its innovative approach, Breakthrough Starshot embodies specific anthropocentric frameworks:

1. **Human timescale concern** prioritizing missions potentially completable within human lifespans
2. **Visual confirmation emphasis** prioritizing imagery comprehensible to human perception
3. **Earth-return communication** assuming continuing human technological civilization
4. **Human-comprehensible data** designing for information interpretable through existing frameworks
5. **Target selection based on Earth-analog potential** rather than genuinely alien environments

These dimensions don't represent limitations but specific choices reflecting human priorities in cosmic exploration. They highlight how even innovative private initiatives remain shaped by human perceptual frameworks and interests rather than attempting to transcend anthropocentric perspectives entirely.

The METI Controversy (Messaging Extraterrestrial Intelligence)

CONCEPT OVERVIEW: METI (Messaging Extraterrestrial Intelligence) involves deliberately transmitting signals into space intended for detection by potential extraterrestrial civilizations, inverting traditional SETI's passive listening approach. Various organizations and individuals have conducted METI transmissions, including:

- The Arecibo Message (1974)
- Cosmic Calls (1999, 2003)
- A Message From Earth (2008)
- Lone Signal project (2013)
- METI International transmissions (2017-present)

Methodological Approaches: METI transmissions typically employ:

1. High-power radio telescopes directing focused signals toward specific stellar targets
2. Messages designed for potential decoding through mathematical or binary principles
3. Content including basic scientific knowledge, human cultural information, and Earth context
4. Transmission durations from minutes to hours
5. Targeting nearby star systems or clusters to maximize reception probability

Controversy Dimensions: METI activities have generated significant scientific controversy around several dimensions:

1. **Existential risk concerns**: Potential danger from hostile or indifferently destructive responses
2. **Representation questions**: Who has authority to speak for humanity?
3. **Consent issues**: Whether unilateral contact attempts require broader societal consent
4. **Signal efficacy debates**: Whether current transmission

methods could plausibly succeed

5. **Ethical uncertainties**: Responsibilities toward potential recipients and their environments

Philosophical Analysis: The METI controversy reveals deeper philosophical questions in extraterrestrial communication:

1. **Risk assessment under uncertainty**: How to evaluate risks with no historical precedent
2. **Collective action problems**: How global commons decisions should be governed
3. **Anthropocentric communication assumptions**: Whether human communication paradigms apply across species boundaries
4. **Temporal ethics**: Responsibilities to future generations potentially affected by contact
5. **Information hazard considerations**: Whether certain knowledge types might be inherently dangerous across civilizations

These philosophical dimensions highlight that METI represents not merely a scientific activity but a profound ethical and political choice with potential intergenerational consequences. The controversy reflects broader tensions regarding technological governance under uncertainty and collective decision-making about actions with speculative but potentially significant consequences.

Current Status: METI activities continue primarily through private organizations, with ongoing debate about appropriate governance and risk assessment frameworks. Notably, no international legal framework specifically regulates METI activities, though proposals for moratoriums or governance structures periodically emerge in scientific and policy communities.

Future Directions in Extraterrestrial Search

The Galileo Project and UAP Investigation

PROGRAM OVERVIEW: Initiated in 2021 by Harvard astronomer Avi Loeb, the Galileo Project represents a scientific approach to investigating Unidentified Aerial Phenomena (UAP) and potential extraterrestrial artifacts. Unlike traditional SETI focused on distant signals or exoplanet biosignatures, the Galileo Project employs systematic observation of Earth's immediate environment for potential technosignatures.

Methodological Approach:

1. Development of automated observatory systems with multiple sensor types (optical, infrared, radar, audio)
2. Systematic data collection protocols with AI-assisted anomaly detection
3. High-resolution tracking capabilities for rapidly moving objects
4. Sensor fusion approach integrating multiple data streams
5. Transparent scientific methodology and public data release

Current Status: The project has established observatories with multi-sensor arrays and begun systematic data collection. While no confirmed extraterrestrial artifacts or phenomena have been reported to date, the project represents an important methodological approach to UAP investigation through transparent scientific protocols rather than classified military observation.

Epistemological Significance: The Galileo Project represents an important epistemological shift in extraterrestrial investigation:

1. **Local environment focus** rather than distant astronomical

observation
2. **Artifact-oriented search** complementing signal or biosignature approaches
3. **Data-first methodology** collecting systematically before developing explanatory frameworks
4. **Multi-sensor integration** acknowledging phenomena might manifest across different detection modalities
5. **Open science model** providing transparent methodology and data access

These approaches address longstanding epistemological problems in UAP investigation, particularly the tension between witness testimony (often from credible observers but inherently subjective) and systematic scientific data collection (objective but typically absent in momentary anomalous events).

Anthropocentric Dimensions: Despite its methodological innovations, the Galileo Project operates within specific anthropocentric frameworks:

1. Focus on visually observable phenomena at human-relevant scales
2. Emphasis on structured objects comprehensible within human manufacturing paradigms
3. Implementation of pattern recognition based on human-defined anomaly categories
4. Target selection reflecting human transportation environment (atmosphere, near-Earth space)
5. Detection technologies optimized for human-recognizable patterns of technological activity

These dimensions reflect practical constraints rather than conceptual limitations, but they highlight the persistent challenges of developing truly agnostic approaches to potential extraterrestrial phenomena that might manifest in ways fundamentally different from human technology or behavior.

Machine Learning Approaches to Anomaly Detection

METHODOLOGICAL INNOVATION: Recent extraterrestrial search programs increasingly incorporate machine learning systems for detecting potential technosignatures or biosignatures that might escape human recognition. These approaches include:

1. **Unsupervised anomaly detection** identifying patterns deviating from baseline natural phenomena
2. **Neural network signal classification** distinguishing potential artificial signals from radio frequency interference
3. **Computer vision analysis** for unusual astronomical imagery or sensor data
4. **Multi-parameter correlation detection** identifying subtle relationships across disparate datasets
5. **Automated feature extraction** from complex spectroscopic or time-series data

Current Applications:

- Breakthrough Listen employing machine learning for radio signal classification
- NASA's Frontier Development Lab developing ML approaches for exoplanet biosignature analysis
- The Galileo Project implementing computer vision for UAP tracking and classification
- Various SETI projects utilizing neural networks for

electromagnetic anomaly detection
- Planetary science missions using ML for identifying unusual geological features

Epistemological Significance: Machine learning approaches offer potential epistemological advantages:

1. **Pattern recognition beyond human perception** revealing relationships not obvious to human observers
2. **Reduced confirmation bias** through algorithm-driven rather than expectation-driven detection
3. **Scalability to massive datasets** beyond human analytical capacity
4. **Novel feature identification** without preconceived classification schemes
5. **Continuous learning** adapting detection parameters based on new information

These advantages suggest ML systems might recognize patterns humans would miss due to perceptual limitations or conceptual biases, potentially expanding detection beyond anthropocentric parameters.

Limitations and Challenges: However, ML approaches face significant limitations:

1. **Training data constraints**: Systems trained on Earth phenomena may fail to recognize truly novel patterns
2. **Algorithmic bias**: ML systems inherit biases present in their design and training processes
3. **Black box problems**: Complex neural networks produce results with limited explanatory transparency
4. **Anomaly contextualization challenges**: Distinguishing

significant anomalies from statistical outliers
5. **Anthropocentric feature selection**: Initial parameter selection still reflects human conceptual frameworks

These limitations reveal that ML approaches don't eliminate anthropocentrism but potentially relocate it from explicit search parameters to implicit algorithm design and training methodologies. ML systems expand but don't transcend the human conceptual frameworks that create them.

Future Directions: Emerging approaches attempt to address these limitations through:

1. **Explainable AI development** creating more transparent reasoning processes
2. **Adversarial training** deliberately challenging systems with edge cases
3. **Minimally biased feature extraction** reducing human parameter selection
4. **Multi-model ensembles** combining different algorithmic approaches
5. **Human-AI collaborative frameworks** integrating complementary capabilities

These developments suggest evolving human-machine collaborative approaches to extraterrestrial search, potentially combining human conceptual creativity with machine perceptual capabilities beyond human limitations.

Interstellar Object Investigation

PHENOMENON OVERVIEW: The 2017 discovery of 'Oumuamua—the first confirmed interstellar object passing through our solar system—created new opportunities for potential extraterrestrial artifact investigation. Subsequently, additional interstellar objects including comet 2I/Borisov have been identified, suggesting regular interstellar visitors potentially observable through improved detection capabilities.

Scientific Approaches: Current and developing methodologies include:

1. **Real-time detection networks** identifying interstellar objects early in their solar system passage
2. **Rapid response observation** deploying multiple telescopes for comprehensive characterization
3. **Spectroscopic analysis** determining composition and potential artificiality markers
4. **Trajectory reconstruction** analyzing motion for natural versus potential technological propulsion
5. **Future interception mission concepts** for direct sampling or observation

'Oumuamua Case Study: 'Oumuamua generated scientific debate regarding potential artificial origin based on several anomalous characteristics:

- Extreme aspect ratio (5-10:1) unlike known natural objects
- Lack of observable coma despite apparent acceleration
- Trajectory deviating from purely gravitational motion
- Unusual brightness variations suggesting extreme shape
- Limited spectroscopic data consistent with manufactured materials

While most astronomers favor natural explanations (likely a fragment from another stellar system), the case highlighted methodological challenges in distinguishing potential artifacts from unusual natural objects, particularly with limited observational data from brief encounters.

Epistemological Significance: Interstellar object investigation reveals important epistemological dimensions:

1. **Limited observation window challenges** requiring rapid methodological deployment
2. **Classification framework limitations** when encountering unprecedented phenomena
3. **Explanatory conservatism tensions** balancing parsimony against genuine novelty
4. **Multi-model comparative analysis** evaluating competing natural and artificial hypotheses
5. **Observation-theory feedback loops** where theoretical frameworks shape observational priorities

These dimensions highlight how extraterrestrial artifact investigation operates at the intersection of observational constraints, explanatory frameworks, and scientific conservatism regarding extraordinary claims. The field demonstrates that artifact search differs fundamentally from SETI signal detection, requiring different epistemological approaches and evidential standards.

Future Developments: Emerging initiatives include:

1. The Vera C. Rubin Observatory potentially detecting numerous additional interstellar objects
2. Mission concepts for rapid deployment to future interstellar visitors
3. Dedicated observation networks specifically targeting interstellar object detection
4. Development of formal protocols for potential artifact

investigation
5. Theoretical modeling to better distinguish unusual natural objects from potential artifacts

These developments suggest a maturing field increasingly capable of systematic investigation rather than opportunistic observation, potentially establishing more robust methodological frameworks for distinguishing natural interstellar objects from hypothetical artifacts.

Conclusion: Methodological Evolution in Extraterrestrial Search

THIS EXAMINATION OF extraterrestrial search programs reveals several overarching patterns in methodological evolution:

1. **Parameter space expansion**: From narrow frequency ranges and target selections to increasingly comprehensive search domains across multiple detection modalities
2. **Anthropocentrism awareness**: Growing recognition of how human cognitive biases and cultural frameworks shape search parameters, driving more deliberate variation in methodological approaches
3. **Epistemological sophistication**: Evolution from binary detection paradigms toward probabilistic frameworks incorporating multiple lines of evidence and degrees of certainty
4. **Technological leverage**: Increasing utilization of artificial intelligence, massive data processing, and sensor fusion to expand detection capabilities beyond human perceptual limitations
5. **Interdisciplinary integration**: Breaking down boundaries between traditional SETI, astrobiology, and planetary

science toward more comprehensive search strategies

These evolutionary patterns don't eliminate the fundamental anthropocentric constraints identified throughout this book but represent increasingly sophisticated engagement with these constraints. Rather than claiming to transcend human perspective entirely, advanced extraterrestrial search methodologies acknowledge their inevitable emergence from human frameworks while systematically varying assumptions to explore wider possibility spaces.

The most sophisticated contemporary approaches embrace what philosopher of science Helen Longino calls "contextual empiricism"—recognition that observation occurs within theoretical contexts while maintaining commitment to empirical constraints on knowledge claims. They develop what historian of science Thomas Kuhn termed "essential tension" between paradigmatic frameworks and genuine openness to anomalous phenomena that might challenge those frameworks.

These methodological developments suggest potentially productive directions for future extraterrestrial search:

1. **Multi-parameter search engines** integrating diverse detection methodologies without privileging single approaches
2. **Alternative intelligence modeling** systematically exploring non-human cognitive architectures and their potential manifestations
3. **Anomaly-centered observation** prioritizing unexplained phenomena before categorization
4. **Agnostic biosignature approaches** developing techniques for recognizing organizational patterns beyond Earth-like

biochemistry

5. **Methodological pluralism** maintaining multiple parallel search strategies rather than converging on consensus approaches

Rather than representing a "solution" to the anthropocentric constraints identified throughout this book, these approaches constitute more sophisticated engagement with these constraints—acknowledging the inevitable shaping of extraterrestrial search by human frameworks while developing practices that might partially transcend their limitations. They suggest an extraterrestrial search characterized by greater epistemological humility, methodological diversity, and openness to phenomena beyond current conception—an approach that might help us recognize truly alien existence if we encounter it, even if it takes forms quite different from those imagined in conventional extraterrestrialist thinking.

Bibliography

Extraterrestrial Search and SETI

Ball, John A. "The Zoo Hypothesis." *Icarus* 19, no. 3 (1973): 347-349.

Billingham, John, and Rudolf Pesek, eds. *The Search for Extraterrestrial Intelligence: SETI.* Cambridge, MA: MIT Press, 1979.

Bostrom, Nick. "Are We Living in a Computer Simulation?" *Philosophical Quarterly* 53, no. 211 (2003): 243-255.

Bracewell, Ronald N. *The Galactic Club: Intelligent Life in Outer Space.* San Francisco: W.H. Freeman, 1975.

Cirkovic, Milan M. *The Great Silence: Science and Philosophy of Fermi's Paradox.* Oxford: Oxford University Press, 2018.

Clarke, Arthur C. *Profiles of the Future: An Inquiry into the Limits of the Possible.* New York: Harper & Row, 1962.

Cocconi, Giuseppe, and Philip Morrison. "Searching for Interstellar Communications." *Nature* 184, no. 4690 (1959): 844-846.

Davies, Paul. *The Eerie Silence: Renewing Our Search for Alien Intelligence.* Boston: Houghton Mifflin Harcourt, 2010.

Dean, Jodi. *Aliens in America: Conspiracy Cultures from Outerspace to Cyberspace.* Ithaca: Cornell University Press, 1998.

Dick, Steven J. *The Biological Universe: The Twentieth Century Extraterrestrial Life Debate and the Limits of Science.* Cambridge: Cambridge University Press, 1996.

Drake, Frank, and Dava Sobel. *Is Anyone Out There? The Scientific Search for Extraterrestrial Intelligence.* New York: Delacorte Press, 1992.

Dyson, Freeman J. "Search for Artificial Stellar Sources of Infrared Radiation." *Science* 131, no. 3414 (1960): 1667-1668.

Finney, Ben R., and Eric M. Jones, eds. *Interstellar Migration and the Human Experience.* Berkeley: University of California Press, 1985.

Hanson, Robin. "The Great Filter - Are We Almost Past It?" Unpublished manuscript, 1998. http://hanson.gmu.edu/greatfilter.html.

Kardashev, Nikolai S. "Transmission of Information by Extraterrestrial Civilizations." *Soviet Astronomy* 8 (1964): 217-221.

Kuhn, Thomas S. *The Structure of Scientific Revolutions.* Chicago: University of Chicago Press, 1962.

Loeb, Avi. *Extraterrestrial: The First Sign of Intelligent Life Beyond Earth.* Boston: Houghton Mifflin Harcourt, 2021.

Sagan, Carl. *Communication with Extraterrestrial Intelligence (CETI).* Cambridge, MA: MIT Press, 1973.

Sagan, Carl, and Frank Drake. "The Search for Extraterrestrial Intelligence." *Scientific American* 232, no. 5 (1975): 80-89.

Shklovskii, I.S., and Carl Sagan. *Intelligent Life in the Universe.* San Francisco: Holden-Day, 1966.

Smart, John M. "The Transcension Hypothesis: Sufficiently Advanced Civilizations Invariably Leave Our Universe." *Acta Astronautica* 78 (2012): 55-68.

Tarter, Jill C. "The Search for Extraterrestrial Intelligence (SETI)." *Annual Review of Astronomy and Astrophysics* 39, no. 1 (2001): 511-548.

Tipler, Frank J. "Extraterrestrial Intelligent Beings Do Not Exist." *Quarterly Journal of the Royal Astronomical Society* 21 (1980): 267-281.

Webb, Stephen. *If the Universe Is Teeming with Aliens... Where Is Everybody? Seventy-Five Solutions to the Fermi Paradox and the Problem of Extraterrestrial Life*. 2nd ed. New York: Springer, 2015.

Astrobiology and Exoplanet Science

BENNER, STEVEN A. "Defining Life." *Astrobiology* 10, no. 10 (2010): 1021-1030.

Chyba, Christopher F., and Kevin P. Hand. "Astrobiology: The Study of the Living Universe." *Annual Review of Astronomy and Astrophysics* 43 (2005): 31-74.

Cleland, Carol E., and Christopher F. Chyba. "Defining 'Life'." *Origins of Life and Evolution of the Biosphere* 32, no. 4 (2002): 387-393.

Des Marais, David J., et al. "The NASA Astrobiology Roadmap." *Astrobiology* 8, no. 4 (2008): 715-730.

Grinspoon, David. *Lonely Planets: The Natural Philosophy of Alien Life*. New York: Ecco, 2003.

Impey, Chris. *The Living Cosmos: Our Search for Life in the Universe.* Cambridge: Cambridge University Press, 2011.

Kasting, James F. *How to Find a Habitable Planet.* Princeton: Princeton University Press, 2010.

Lovelock, James E. *Gaia: A New Look at Life on Earth.* Oxford: Oxford University Press, 1979.

Margulis, Lynn. *Symbiotic Planet: A New Look at Evolution.* New York: Basic Books, 1998.

Rothschild, Lynn J., and Rocco L. Mancinelli. "Life in Extreme Environments." *Nature* 409, no. 6823 (2001): 1092-1101.

Seager, Sara. *Exoplanet Atmospheres: Physical Processes.* Princeton: Princeton University Press, 2010.

Shapiro, Robert. *Origins: A Skeptic's Guide to the Creation of Life on Earth.* New York: Summit Books, 1986.

Ward, Peter D., and Donald Brownlee. *Rare Earth: Why Complex Life Is Uncommon in the Universe.* New York: Copernicus, 2000.

Wordsworth, Robin. "Atmospheric Evolution on Inhabited and Lifeless Worlds." *Science* 368, no. 6495 (2020): 1055-1056.

Philosophy of Science and Epistemology

BACHELARD, GASTON. *The Formation of the Scientific Mind.* Manchester: Clinamen Press, 2002.

Bogost, Ian. *Alien Phenomenology, or What It's Like to Be a Thing.* Minneapolis: University of Minnesota Press, 2012.

Braidotti, Rosi. *The Posthuman.* Cambridge: Polity Press, 2013.

Cartwright, Nancy. *The Dappled World: A Study of the Boundaries of Science*. Cambridge: Cambridge University Press, 1999.

Haraway, Donna J. "Situated Knowledges: The Science Question in Feminism and the Privilege of Partial Perspective." *Feminist Studies* 14, no. 3 (1988): 575-599.

Harman, Graham. *Object-Oriented Ontology: A New Theory of Everything*. London: Pelican Books, 2018.

Harding, Sandra. *The Science Question in Feminism*. Ithaca: Cornell University Press, 1986.

Kohn, Eduardo. *How Forests Think: Toward an Anthropology Beyond the Human*. Berkeley: University of California Press, 2013.

Kuhn, Thomas S. *The Structure of Scientific Revolutions*. Chicago: University of Chicago Press, 1962.

Latour, Bruno. *We Have Never Been Modern*. Cambridge, MA: Harvard University Press, 1993.

Meillassoux, Quentin. *After Finitude: An Essay on the Necessity of Contingency*. London: Continuum, 2008.

Morton, Timothy. *Hyperobjects: Philosophy and Ecology after the End of the World*. Minneapolis: University of Minnesota Press, 2013.

Nagel, Thomas. "What Is It Like to Be a Bat?" *The Philosophical Review* 83, no. 4 (1974): 435-450.

Popper, Karl. *The Logic of Scientific Discovery*. London: Routledge, 2002.

Stengers, Isabelle. *Cosmopolitics*. Minneapolis: University of Minnesota Press, 2010.

Tsing, Anna Lowenhaupt. *The Mushroom at the End of the World: On the Possibility of Life in Capitalist Ruins*. Princeton: Princeton University Press, 2015.

Whitehead, Alfred North. *Process and Reality*. New York: Free Press, 1978.

Wittgenstein, Ludwig. *Philosophical Investigations*. Oxford: Blackwell, 1953.

Wolfe, Cary. *What Is Posthumanism?* Minneapolis: University of Minnesota Press, 2010.

Anthropology and Cultural Studies

ABRAM, DAVID. *The Spell of the Sensuous: Perception and Language in a More-Than-Human World*. New York: Pantheon Books, 1996.

Battaglia, Debbora, ed. *E.T. Culture: Anthropology in Outerspaces*. Durham: Duke University Press, 2005.

Descola, Philippe. *Beyond Nature and Culture*. Chicago: University of Chicago Press, 2013.

Devereux, Paul. *Earth Lights Revelation: UFOs and Mystery Lightform Phenomena: The Earth's Secret Energy Force*. London: Blandford, 1989.

Hallowell, A. Irving. "Ojibwa Ontology, Behavior, and World View." In *Culture in History: Essays in Honor of Paul Radin*, edited by Stanley Diamond, 19-52. New York: Columbia University Press, 1960.

Helmreich, Stefan. *Alien Ocean: Anthropological Voyages in Microbial Seas*. Berkeley: University of California Press, 2009.

Kohn, Eduardo. *How Forests Think: Toward an Anthropology Beyond the Human*. Berkeley: University of California Press, 2013.

Lepselter, Susan. *The Resonance of Unseen Things: Poetics, Power, Captivity, and UFOs in the American Uncanny*. Ann Arbor: University of Michigan Press, 2016.

Mack, John E. *Abduction: Human Encounters with Aliens*. New York: Charles Scribner's Sons, 1994.

Oreskes, Naomi, and Erik M. Conway. *Merchants of Doubt: How a Handful of Scientists Obscured the Truth on Issues from Tobacco Smoke to Global Warming*. New York: Bloomsbury Press, 2010.

Said, Edward W. *Orientalism*. New York: Pantheon Books, 1978.

Smith, Cameron M. *Principles of Space Anthropology: Establishing a Science of Human Space Settlement*. Cham: Springer, 2019.

Taussig, Michael. *Mimesis and Alterity: A Particular History of the Senses*. New York: Routledge, 1993.

Tsing, Anna Lowenhaupt. *The Mushroom at the End of the World: On the Possibility of Life in Capitalist Ruins*. Princeton: Princeton University Press, 2015.

Viveiros de Castro, Eduardo. *Cannibal Metaphysics*. Minneapolis: Univocal Publishing, 2014.

Physics and Space Exploration

BARROW, JOHN D., AND Frank J. Tipler. *The Anthropic Cosmological Principle*. Oxford: Oxford University Press, 1986.

Bohm, David. *Wholeness and the Implicate Order*. London: Routledge & Kegan Paul, 1980.

Cameron Smith, Cameron. *Principles of Space Anthropology: Establishing a Science of Human Space Settlement*. Cham: Springer, 2019.

Davies, Paul. *The Fifth Miracle: The Search for the Origin and Meaning of Life*. New York: Simon & Schuster, 1999.

Dyson, Freeman. *Disturbing the Universe*. New York: Harper & Row, 1979.

Hameroff, Stuart, and Roger Penrose. "Consciousness in the Universe: A Review of the 'Orch OR' Theory." *Physics of Life Reviews* 11, no. 1 (2014): 39-78.

Ho, Mae-Wan. *The Rainbow and the Worm: The Physics of Organisms*. Singapore: World Scientific, 2008.

Lim, Eugene. "Searching for Dark Matter Life." Unpublished manuscript, 2020. arXiv:2012.02799.

Randall, Lisa. *Dark Matter and the Dinosaurs: The Astounding Interconnectedness of the Universe*. New York: Ecco, 2015.

Rees, Martin J. *Before the Beginning: Our Universe and Others*. Reading, MA: Addison-Wesley, 1997.

Sagan, Carl. *Pale Blue Dot: A Vision of the Human Future in Space*. New York: Random House, 1994.

Tsiolkovsky, Konstantin. "The Exploration of Cosmic Space by Means of Reaction Devices." *The Science Review* 5 (1903).

Religious Studies and Theology

BARRETT, JUSTIN L. *Why Would Anyone Believe in God?* Walnut Creek, CA: AltaMira Press, 2004.

Däniken, Erich von. *Chariots of the Gods?* New York: Putnam, 1970.

de Chardin, Pierre Teilhard. *The Phenomenon of Man.* New York: Harper & Row, 1959.

Davies, Paul. *The Mind of God: The Scientific Basis for a Rational World.* New York: Simon & Schuster, 1992.

Lewis, C.S. "Religion and Rocketry." In *The World's Last Night and Other Essays*, 83-92. New York: Harcourt, Brace & World, 1960.

Peters, Ted. *God as Trinity: Relationality and Temporality in Divine Life.* Louisville: Westminster/John Knox Press, 1993.

Roberts, Mere. "Indigenous Knowledge and the Development of Western Science." PhD diss., University of Cambridge, 1998.

Sitchin, Zecharia. *The 12th Planet.* New York: Stein and Day, 1976.

Tempier, Étienne. "The Condemnation of 219 Propositions." In *Medieval Political Philosophy: A Sourcebook*, edited by Ralph Lerner and Muhsin Mahdi, 335-354. Ithaca: Cornell University Press, 1963.

Waltemathe, Michael. "Bridging Multiple Realities: Religion, Play and Alfred Schutz's Theory of the Life-World." In *Playing with Religion in Digital Games*, edited by Heidi A. Campbell and Gregory P. Grieve, 238-254. Bloomington: Indiana University Press, 2014.

Wilkinson, David. *Science, Religion, and the Search for Extraterrestrial Intelligence.* Oxford: Oxford University Press, 2013.

General Works

BRIN, DAVID. *The Transparent Society: Will Technology Force Us to Choose Between Privacy and Freedom?* Reading, MA: Addison-Wesley, 1998.

Bryson, Bill. *A Short History of Nearly Everything.* New York: Broadway Books, 2003.

Chiang, Ted. *Stories of Your Life and Others.* New York: Tor Books, 2002.

Clarke, Arthur C. *2001: A Space Odyssey.* New York: New American Library, 1968.

Dick, Steven J. *Plurality of Worlds: The Origins of the Extraterrestrial Life Debate from Democritus to Kant.* Cambridge: Cambridge University Press, 1982.

Goldstein, Rebecca Newberger. *Incompleteness: The Proof and Paradox of Kurt Gödel.* New York: W.W. Norton, 2005.

Le Guin, Ursula K. *The Left Hand of Darkness.* New York: Ace Books, 1969.

Lem, Stanisław. *Solaris.* New York: Walker, 1970.

Marcel, Gabriel. *Being and Having.* Westminster: Dacre Press, 1949.

Paulos, John Allen. *Innumeracy: Mathematical Illiteracy and Its Consequences.* New York: Hill and Wang, 1988.

Sagan, Carl. *Cosmos.* New York: Random House, 1980.

Strugatsky, Arkady, and Boris Strugatsky. *Roadside Picnic.* New York: Macmillan, 1977.

Wells, H.G. *The War of the Worlds*. London: William Heinemann, 1898.